DECISION TABLES

DECISION TABLES

Michael
Montalbano

SCIENCE RESEARCH ASSOCIATES, INC.
Chicago, Palo Alto, Toronto, Henley-on-Thames, Sydney, Paris
A Subsidiary of IBM

© 1974 Science Research Associates, Inc.
All rights reserved.

Printed in the United States of America
Library of Congress Catalog Card Number 73-92630
ISBN 0-574-19040-6

To the memory of my parents

Pietro Montalbano
born, Bagheria, Sicily, April 6, 1885
died, Brentwood, New York, September 29, 1962

Rosaria Di Stephano Montalbano
born, Cefalú, Sicily, December 9, 1885
died, Brentwood, New York, July 13, 1971

Preface

By the time he starts to think about a preface, the writer of a book is more intimately and painfully aware of its shortcomings than any reader will ever be. The preface is his last chance to say something about the failings that concern him most. In this preface I'd like to discuss three aspects of this book: its moral, its origins, and its intended audience.

By *moral* I mean what Aesop meant—a short, pithy statement of point or message. The moral of this book is

> *If you wish to use digital computers effectively, the first thing you should do is digitize your procedure descriptions.*

By *digitize*, I mean "put in digital form." This is what we mean when we speak of digitizing a photograph. A photograph is a pictorial description of some object of interest—a bubble-chamber track, a lunar crater, a cloud formation. If we wish to use a digital computer to process the information in a photograph, we must convert the pictorial description into a digital description—a sequence of digits whose structure is known. Physical devices called digitizers are used to make this conversion.

Unfortunately, we do not have physical digitizers to translate from conventional procedure descriptions to the digital descriptions required by a computer. What we do have is the beginnings of a conceptual digitizer: the decision table. The objective of this book is to introduce the reader to the decision table as a tool for putting procedure descriptions in digital form—that is, "digitizing" them.

The ideas expounded in this book are a limited, specific expression of a general attitude about data processing that has been formed by a variety of experiences over many years. I first started programming digital computers before any were in existence, in May 1948, at the National Bureau of Standards in Washington. This was a year before the first stored-program digital computer (the EDSAC, in England) was completed, and two years before completion of the first American computer (the SEAC, built by the National Bureau of Standards). The only programming possible was for "paper" machines—machines proposed by various vendors to such government agencies as the Bureau of the Census, the Army Map Service, and the Office of the Air Comptroller.

Despite the fact that the machines were not yet in existence, it was evident even then that the engineers were ahead of the programmers, that technology was ahead of methodology, that we could produce gadgets faster than we could produce ideas. My subsequent experience—at the Naval Research Laboratory, the Kaiser Steel Corporation, and IBM—has strengthened my early conviction that the most pressing problems facing us are in the area of methodology and design.

The first form of the "branch label" idea was developed in 1956, while I was employed by Kaiser Steel, as my contribution to a Business Language Research Project conducted jointly by Kaiser Steel, IBM, and Touche, Niven, Bailey and Smart. The idea that procedure descriptions should be digitized was expressed in a paper, "Winnowing Information from Data," that I gave at a seminar on data processing and management information held at MIT during the summer of 1957. From 1962 through 1967 I had the opportunity to engage in research and teaching in the field of business information systems at the Graduate School of Business of Stanford University. Many of the ideas disucssed in the book were used in student projects undertaken either as sponsored research or as class assignments.

I hope the book will be of interest to anyone concerned with the efficient use of digital computers. In particular, I hope it will be useful to professionals of three types: systems analysts, programmers, and computer scientists.

I have described the plight of the programmer. The inefficiences of our present way of doing things make themselves most painfully evident in the form of programming costs and delays. Yet appearances are misleading. Programming difficulties result from poor design, from our unwillingness to systematize the work of the systems analyst. Programmers are under the gun merely because they are the ones with the final deadlines; others can slip their schedules, programmers cannot. The early slippages are not as noticeable because the output of those with the early deadlines—the systems analysts—is not as clean-cut or definite as a program. Until it is, the programmer will continue to be plagued by problems not of his own making. For the programmer, the decision table offers the hope that he will get clear, complete, consistent, timely program specifications, a luxury that at present is all too frequently denied him.

I do not mean to imply that the systems analyst has a bed of roses. The programmer, because he is conspicuous as a bottleneck, has been provided a wide variety of tools; but no such assistance has been provided the systems analyst. The programmer is primarily concerned with machines; the systems analyst is primarily concerned with people. Machines are a good deal more tractable than people, and machine tools are easier to provide than people tools. For the systems analyst, then, the decision table can be a start toward the tool kit that will eventually permit him to introduce order, regularity, and intelligibility into the chaotic world in which he now lives.

If this is to happen, it will require the cooperation of programmer, systems analyst, and computer scientist. The task of the computer scientist is to do the research required to increase the power and utility of computers. All too often, the easiest avenues of computer science research are those which convert practical computer problems into exercises in pure mathematics—topology, say, or logic. Unfortunately, the most pressing problems are those represented by the messiest kind of impure mathematics—the mathematics of efficiency rather than elegance. The pressing need is not for existence theorems, but for ingenious algorithms, programs, procedures, techniques, and ideas that will help solve the costly and difficult problems that confront the day laborers in the data-processing vineyard. These problems are, or should be, of enormous theoretical interest precisely because no satisfactory theory about them now exists. It would make me very happy if this book aroused the interest of the computer science students by whom such a theory has to be developed if it is to be developed at all.

Palo Alto, California
December 18, 1973

Contents

1	The Setting	1
2	Structure and Use	7
3	Narrative Descriptions and Flowcharts	31
4	Simple Rules and Composite Rules	49
5	Completeness and Consistency I	65
6	Completeness and Consistency II	81
7	The Algebra and Geometry of Logic	97
8	Decision Tables and Flowcharts	129
9	Decision-Table Translators	147
10	Open Questions and Opinions	155
	Answers to Exercises	165
	Index	189

DECISION TABLES

1
The Setting

A. INTRODUCTION: FACTS AND OPINIONS

The facts about decision tables are few; most of them are presented and discussed briefly in chapter 2. The remainder of this book is an attempt to flesh out these facts by discussing them at some length and illustrating where, when, and how they apply to practical problems in the field of data processing.

This means that most of this book is opinion. This is true of most books, but the point needs stressing; many books present highly debatable opinion as generally accepted fact. For example, an author will tell us that his favorite programming language is "simple" and leave us to guess what he means by simplicity. Or he "demonstrates" the alleged simplicity by presenting statistics on how long it takes the "average student" to learn how to write a "significant" program. Now we are left to guess what he means by "average student" and "significant" and what he thinks his statistics have "demonstrated." All too frequently, this masquerading of opinion as fact disfigures the polemics of a field in which there are few objectively demonstrable facts but many passionately held opinions. For example, when time-sharing systems were first being developed, one of the originators of a particular system asserted that it had already justified itself on the basis of the "research that has been done that would not otherwise have been done." He did not mention what the research was, nor what it had cost. Didn't he think it relevant to compare the value of what had been done against the expenditure it had required? Didn't he think it proper to offset the value of what had been done by the value of what could have been done but was *not* done because the researcher was mesmerized by a new toy?

This, of course, is no place to discuss questions like these. This book is about decision tables. The only opinions it is concerned with are those which people hold about decision tables. It is not surprising to find that these opinions are not only about whether decision tables are "good" or "bad" but even, among those who are convinced they are good, what it is they are good for and how and where they can be used most effectively. Here, for example, is a list of some of the opinions expressed by proponents of decision tables:

1. They are superior to narrative descriptions or flowcharts, because they are easier to construct, easier to understand, and easier to modify.
2. Their structure encourages the user to break his description of a data-processing procedure into interconnected "modules," each of which is relatively independent of the others.
3. Decision-table translators are useful, particularly those which produce programs in a widely used programming language such as COBOL, FORTRAN, or RPG.
4. The limited-entry form of decision table is the most useful.
5. The present form of decision table requires little further development.

The author's opinions on these questions will be presented, explicitly or implicitly, throughout this book. The impatient reader can find them summarized in the concluding chapter. It would be greatly preferable, however, if he would read what follows in the order in which it is presented and judge for himself both what he guesses the author's biases to be and what the actual merits of decision tables are.

The fundamental problems with which decision tables are concerned are those of understanding, analyzing, designing, and describing complex procedures.

- By *understanding*, we mean learning what a procedure is all about.
- By *analyzing*, we mean breaking it down into a set of simple, self-contained procedures or steps.
- By *designing*, we mean constructing a procedure that will achieve specified objectives.
- By *describing*, we mean putting the results of our analysis or design in a form that will be readily comprehensible to others.

The tasks of understanding, analyzing, designing, and describing procedures are, in many organizations, assigned to an employee called a systems analyst. As a way to determine where decision tables are likely to be found useful, it might be well to consider briefly what a systems analyst does.

B. THE NATURE OF SYSTEMS ANALYSIS

A systems analyst is usually a nonspecialist who is required to understand and restructure the activities of specialists in such a way as to achieve new objectives or to achieve old objectives more efficiently. At its inception, his typical assignment is usually a broad, general one such as "Determine how the payroll (or accounts payable, sales analysis, production planning, order entry . . .) procedure can be improved and which portions of it can be mechanized" or, on an even grander scale, "Design a management information system that will tell the management of this company *what it needs to know.*"

This general assignment has to be broken down into a set of more specific ones. This requires a joint effort by both the systems analyst and the specialists from whom he has to gather information and to whom he has to communicate the results of his analysis and design. The process usually starts with a disorganized casting about for relevant information. If successful, it ends with a highly organized definition of the area of application and a concise, complete, consistent description of the specific actions that are to be performed by people or machines or both.

The activities that take place between the first investigations and the final results are complicated and are apt to vary from one problem to another and from one set of people to another. In general, however, we can say that a certain set of steps is repeated:

1. Information is gathered.
2. It is examined to see if it is intelligible, if it describes the application or problem in sufficient detail to permit analysis and design, if it describes all the actions that have to be taken, if it describes all the information that is required before action can be taken, if it contradicts itself, if it repeats itself, and so on.
3. If the gathered information is unsatisfactory for any reason, step 1 is repeated.

The process we have outlined above is beset with many difficulties. The systems analyst finds himself talking to specialists who express themselves in an incomprehensible jargon—the technical terms used in their areas of specialization. These specialists are apt to be busy people who have little time or inclination to discuss their activities. Most of them find it hard to verbalize the "feel" that guides their actions on the job, the day-to-day reflex actions they have learned through long experience. Some parts of their jobs are, to them, so self-evident that they require no explicit statement. They find it hard to understand that they are not quite so evident to others.

In overcoming these difficulties, the systems analyst is required to exercise tact, intelligence, patience, adaptability, and a capacity for disciplined, objective thought. These are the attributes required of anyone who wishes to understand or design a complicated procedure in any field of endeavor, whether the field is science, engineering, business, government, or any other structured activity. The chief advantage that science and engineering possess over the others is that they can avail themselves of an efficient conceptual tool which facilitates the tasks of understanding, analyzing, designing, and describing. This conceptual tool is mathematics, either the simple mathematics of adding, subtracting, multiplying, and dividing or such advanced specialties as numerical analysis, differential equations, algebra, statistical theory, and so on.

At present, the systems analyst who is concerned with the procedures of management has no formal tool like mathematics to assist him in his daily work. The processes he has to understand and improve are less orderly than those occurring in the technical fields. They involve many more individual variables, and these variables interact in much more complicated ways than those which can be treated by the mathematics of the scientist or engineer. What the systems analyst needs is a conceptual tool that has not yet been developed, a tool that will do for him what mathematics does for the technician. At present, the chief tools of his trade are glossaries, narrative descriptions, flowcharts of one kind or another, and a variety of special techniques loosely grouped under the heading "industrial engineering." In the opinion of the author, decision tables constitute a first attempt to develop a conceptual tool for the use of the systems analyst.

One last point in this connection. In most modern organizations, the end product of a systems analyst's efforts is a procedure description, all or part of which has to be communicated to the systems analyst's counterpart—the programmer. As we shall see later in the book, a program is merely a procedure written for computers rather than for people. The programmer's job is to translate the systems analyst's procedure description into a program that will make efficient use of his company's data-processing system.

From the programmer's standpoint, the decision-table form of procedure description is doubly useful. At present, the most difficult part of his job is not the programming task itself. Rather, it is determining *what* to program—that is, understanding the procedure description, or programming specifications, given him by the systems analyst. At present, most of the initial programming specifications he gets are incomplete, inconsistent, and unclear. A great deal of give-and-take between him and the systems analyst is required before these omissions, inconsistencies, and ambiguities can be resolved. The decision-table form of procedure description can eliminate or reduce the time and expense of converting a procedure description into a good set of programming specifications. In addition, since a programmer is a systems analyst whose procedures are for computers rather than people, the decision table provides a conceptual tool for him as well—a tool that will help him design an efficient computer procedure or program. Further, the nature of the tool is such that it makes it possible to mechanize much of the work of both systems analyst and programmer, thus eliminating a great deal of the drudgery that currently characterizes both professions.

C. HOW TO USE THIS BOOK: OBSERVING AND DOING

Most of what we learn is learned neither from teachers nor from books. This does not mean that teachers and books are not needed. It means that if one is to learn from them, he is apt to learn most effectively if he tries to learn in the same way he learns from other educational experiences.

Consider something that every child learns long before his formal education starts: how to tell a dog from a cat, or a fox, or a bear, or a wolf. The process is a mysterious one and it is not clear that anyone understands it fully. Dogs come in a bewildering variety of shapes, sizes, colors, stances, and other attributes. Yet a child quickly learns to recognize that a chihuahua and a St. Bernard are similar to each other in ways in which they are different from other animals. He can even distinguish between a German shepherd and a wolf. No one has ever given him a course in dog identification. How has he learned to recognize a dog?

His fundamental experiences are the association of the word *dog* with a variety of animals. As he considers these animals, he abstracts the qualities that he finds common to all the animals to which the word *dog* is applied and then generalizes these qualities so that they constitute, in his mind, the qualities that define what the word *dog* means. After he has seen a few varieties of dog, his understanding of what *dog* means is good enough to enable him to identify as *dog* a breed that he has never seen before.

This same process of abstracting and generalizing should be applied in more formal kinds of education. The fundamental requisite for learning is the student's own perceptions. Teachers and books can guide or inspire his perceptions, but they cannot substitute for them. This book is one of several about decision tables. In all of the books, examples of decision tables are given. The examples and some of the details of construction will vary with the author. But *decision table,* if it means anything, means what all the examples have in common.

Thus the first principle to follow in reading this or any other instructional material is not merely to *see* but to *observe,* to think about what you are seeing. The ability to abstract comes from *observing*—that is, from active rather than passive seeing.

Merely observing, however, is not sufficient. The objective of this book is to teach the reader how to construct his own decision tables. This objective will not be achieved if he does not, in fact, attempt to construct decision tables. Even if the reader's personal objective is to understand decision tables without necessarily constructing any, he will find constructing decision tables an important aid to understanding. Only if he attempts such construction will he be able to determine whether he has correctly interpreted what he has read, and, more important, only then will he be able to determine for himself whether and where he can make good use of decision tables.

Most of the chapters in the book have illustrations to which the text refers. To facilitate the processes of *observing* and *doing,* all of these are placed at the end of the chapter in which they are discussed. Most of the chapters have a review and exercises as well. It is recommended that the reader study the illustrations, the review, and the exercises *before* he reads the chapter. Learning—like seeing—should be active rather than passive; it should be acquiring, testing, and exercising one's own opinions rather than ingesting the opinions of someone else.

At this point, the reader should attempt the kind of active learning we have just described. Look ahead to the figures, review, and exercises of chapter 2. There are nine figures at the end of the chapter. Consider them one at a time and then as a group. Even before you read about them, can you get an idea of what decision tables are? Can you deduce what all the illustrations have in common? Can you see differences from one illustration to the next and figure out what characteristics each different form of decision table has and under what circumstances one form is preferable to another? Do the illustrations suggest any questions? Does the review establish a framework for the chapter? Do the exercises help suggest what key ideas the chapter will try to communicate? Can you do any of the exercises before you read the chapter? If not, look ahead to the answers if you wish; they are an important part of the book's message. They are intended not as obstacles but as aids. For a subject like decision tables, the worked example is one of the most effective of all teaching and learning tools. Don't hesitate to use the answers at the back of the book as a collection of worked examples if that's the way you learn best.

The organization of chapter 2 and most of the following chapters reflects the author's belief that a reader will find it helpful to form a preliminary idea of each chapter's contents and objectives before he reads the detailed discussion. This is particularly true with a topic like decision tables, where many of the important ideas and facts are primarily about structure; examining and then constructing typical structures is one of the best ways to learn what they are and what they can do.

If the reader uses the summary material at the end of the chapter—the figures, review, and exercises—for both preview and review, he will, on his own, be adopting the pedagogic technique that is often encapsulated in the following advice to a would-be teacher:

1. Tell your students what you're about to tell them.
2. Tell them.
3. Tell them what you told them.

Put in the form of advice to a reader, this might be rephrased as:

1. Read about what you're going to read.
2. Read it.
3. Read about what you read.

The chapters of this book have been organized in such a way as to make this previewing, viewing, and reviewing as effective as possible.

2

Structure and Use

A. INTRODUCTION

Five of the figures at the end of this chapter (figures 2.9 and 2.1 through 2.4) show different ways to describe the same procedure—an artificial one covering the sale of three types of vintage used car. Their purpose is to illustrate different decision-table format conventions. Figures 2.5 and 2.6 display the inner structure of decision tables and introduce the terms used to name their components. Figure 2.7 is a sample form, of which a functional decision table is the major part. The remainder of the form provides information that identifies the table and ties it into other parts of a data-processing system. Figure 2.8 is a description of an income tax procedure.

These figures, and the discussion in the remainder of this chapter, give virtually all the basic facts about decision tables. These facts are few. The fundamental idea is a simple one. So is the use of letters to represent variables in algebra. But, like algebra, the simple idea has important, sometimes complex, implications. The objective of the chapters following this one is to explore some of those implications.

B. DECISION-TABLE FORMATS AND NOMENCLATURE

Formats

The procedure described in figures 2.1 through 2.4 can be put into words in many ways. This is one way:

1. If you sell a Cord or Reo that's running well, your commission is 5 percent.

2. If you sell a Cord that's in poor condition, your commission is 10 percent and you should schedule one week of shopwork to get it in running order.
3. If you sell a Cord or Reo that's not running, your commission is 10 percent and you should schedule six weeks of shopwork to get it in running order.
4. If you sell a Reo that's in poor condition, your commission is 10 percent and you should schedule two weeks of shopwork to get it in running order.
5. Check with the manager before completing any Duesenberg sale to get his permission, to determine what commission you will earn, and to determine how much shopwork will be required to put the car in running order.

The procedure is, of course, artificial—as artificial as the author could devise. Its objective is to illustrate the various forms a decision table can have and to provide a contrast between the decision-table description of a procedure and alternative forms of description, such as the verbal form illustrated above.

The essential difference is one of *structure*. A decision table is divided into parts that serve different purposes. Each decision table is divided into four major parts by intersecting vertical and horizontal double lines. Each of these parts has its own specific function. Two of these parts, for example, are designed to answer the following questions:

1. What actions are governed by this procedure?
2. What factors have to be considered before a course of action can be determined?

In a decision table, the answer to the first of these questions is to be found in the lower left-hand section of the table: the *action stub*. The answer to the second is to be found in the upper left-hand corner: the *condition stub*. In the action stubs of figures 2.1 and 2.3, we find that the actions governed by the procedure they describe are identified as

1. Commission
2. Shopwork
3. Manager O.K.

These three action descriptions are shorthand for the directives:

1. Your commission is . . .
2. The shopwork required is . . .
3. Manager approval (is, is not) required.

They are thus general specifications of the *types* of action governed by the procedure. The complete specification of actual actions is to be found elsewhere in the table. In general, the specific actions to be performed will depend on the conditions governing an actual sale.

Similarly, the condition stub (the upper left-hand corner) contains a description of the types of condition to be considered. For tables 2.1 and 2.3 we find the conditions to be

1. Make
2. Condition

These are shorthand for the questions

1. What is the make?
2. What is the condition?

Only the questions themselves are to be found in the condition stub. Answers are to be found elsewhere in the table.

We are now in a position to compare parts of decision tables 2.1 and 2.3 with the verbal description of the vintage-car procedure given earlier. In the reader's own estimation, how easy would it be—even for a procedure as simple as this one—to determine from the verbal description that only two factors are to be considered in this procedure (make and condition) and that three actions are governed by it (determining commission, estimating shopwork, and obtaining manager approval)?

When a procedure is expressed in words, we usually cannot be sure until we have read the entire description—and made notes on what we have read—that we know everything that has to be considered and everything that has to be done. In point 1, for example, we learn immediately that we have to consider the make and condition of the car we are selling. But it is not until we have finished reading point 5 that we can be *sure* that these two factors are all we have to consider. In point 1, the only action mentioned is *determining commission.* Point 2 informs us that we may also have to *estimate shopwork.* But it is not until we get to point 5 that we learn that we may also have to get *manager approval* of a sale.

The contrast is between form and formlessness. In the decision table we know exactly where to look to find the answers to our two questions. In the verbal—frequently called the *narrative*—description there are no specific places to look; the entire description must be read and analyzed. And this contrast, which is striking even in our simple, artificial example, is much more pronounced in the complicated, intricate procedures that occur in real organizations.

Having noted this important difference between most narrative procedure descriptions and corresponding decision-table descriptions, we should also note that the difference is not truly fundamental. There is no reason why a verbal description could not be given more structure than the one we have chosen as an illustration. In fact, many of the procedure descriptions prepared by good systems-and-procedures or industrial engineering departments undoubtedly do have sections corresponding closely in both form and function to the portions of the decision table called condition stub and action stub. But it is by no means the general rule.

The truly fundamental difference between decision tables and other forms of procedure description, however, is not merely that decision tables have sections called *condition stubs* and *action stubs,* but that these sections are related to each other and to other sections of the table in certain specified, standardized ways.

Before we discuss these relationships, let us examine the stubs of figures 2.1 to 2.4 to see the different forms stubs can have.

Although all four tables describe the same procedure, there is a striking difference between the stubs of figures 2.1 and 2.3 and those of figures 2.2 and 2.4. In the former, the conditions and actions are specified relatively broadly; in the latter, extremely narrowly. In the former, they are closer to the forms of normal conversation; in the latter, they are more like permissible questions in the game "Twenty Questions"—that is, the conditions and actions are described in the form of yes-no questions or do-don't do directions.

The yes-no form of organization is characteristic of what is called a *limited-entry decision table.* In such a table, a question like "What is the make of the car?" must be broken down into a set of further questions: "Is the car a Cord?"; "Is the car a Reo?"; "Is the car a Duesenberg?" (and so on, if more cars are to be covered by the procedure). Similarly, the single action covered by the word *commission* in figures 2.1 and 2.3 is translated into the three actions—"Commission is 5 percent"; "Commission is 10 percent"; "Commission is variable"—in figures 2.2 and 2.4, the *limited-entry* tables.

Figures 2.1 and 2.3, in which conditions and actions are described more broadly, are examples of *extended-entry* tables.

The word *entry,* which forms part of both decision-table descriptions, refers to the two sections of a decision table that we have not yet discussed—the sections to the right of the vertical double lines. These sections are known as *entries:* the one above the horizontal double line is called the *condition entry;* the one below is called the *action entry.* The condition entry contains answers to the questions asked in the condition stub; the action entry contains specifications of the actions described in the action stub. The term *limited-entry* means that the answers in the condition entry are *limited* to *yes* and *no* (or some coded form thereof), and the specifications in the action entry are *limited* to *do-don't do, yes-no,* or some equivalent coded form. When more general answers are permitted, the table is called *extended entry.* Combinations of the two are sometimes called *mixed entry.*

The entry sections are what tie the decision table together. They provide the truly fundamental difference between decision tables and other forms of procedure description.

The difference is in the way the branching structure—commonly called the *program logic*—of a procedure is displayed. Virtually all significant procedures include conditional actions—actions that are performed only if a certain condition or set of conditions holds true. Program logic is this set of interrelationships between conditions and actions. In the most common methods of describing program logic—narrative descriptions and flowcharts—conditions and actions are intermixed. In decision tables, on the other hand, they are separated: all the conditions are to be found in the condition entry; all the actions in the action entry.

The next chapter discusses and compares these alternative methods of describing program logic. Here our purpose is to understand how the condition-entry and action-entry portions of the decision table are constructed and how they are related to each other and to the other sections of the table.

A condition table is divided into *rows* by parallel horizontal lines. The rows above the double line are called *condition rows;* the rows below it are called *action rows*.

The right-hand, or entry, half of a decision table is divided by vertical lines into columns called *rules*. (Numbers identifying the rules appear at the tops of the columns.) Rules traverse both the condition entry and the action entry portions of a table, thus connecting one to the other. The portion of a rule in the condition entry is called the *condition* portion of the rule; the portion in the action entry is the *action* portion.

The condition portion of a rule contains answers to the questions in the condition stub, the answer on any row corresponding to the question on that row. The action portion of a rule contains specifications of the actions described in the action stub, the specification on any row corresponding to the action described on the corresponding row in the action stub.

A rule is, functionally, a conditional directive of the form "If the conditions specified in the condition portion hold, then take the actions specified in the action portion." For example, rule 1 in figure 2.1 can be translated into words as "*If* the car is a Cord that is running well, *then* the commission to be taken on the sale is 5 percent; no shopwork is required; no manager O.K. is required." The same rule, in figure 2.2, can be read as "*If* the car is a Cord and it is not a Reo and it is not a Duesenberg and it is running well and it is not running poorly and it is not not running, *then* commission is 5 percent, not 10 percent and not variable; no shopwork is required, one week of shopwork is not required, two weeks of shopwork are not required, six weeks of shopwork are not required, no estimate of shopwork is required; no manager O.K. is required."

The difference between the verbal statements of equivalent rules in the extended-entry and limited-entry tables suggests something about the comparative structure of the two which will be discussed more fully later in the book. Meanwhile the reader is urged to make his own comparison.

We have called figures 2.3 and 2.4 *coded* versions of figures 2.1 and 2.2. Coding is an operation fundamental to all of data processing. We can use digital computers effectively only when we have devised effective codes for the objects and relationships with which they are to deal.

When we code, we make a correspondence between a wieldy set of symbols (the code) and an unwieldy set, usually consisting of words or other relatively informal representations of objects, actions, conditions, and so on.

In the limited-entry decision table, the only things we need code are symbols for *yes* and *no,* or *do* and *don't do.* The customary convention is to code a 1 for *yes* or *do* and a 0 for *no* or *don't do.* This coding is so simple and so common that it is often assumed rather than explicitly stated.

In an extended-entry table, on the other hand, the meaning of a particular code depends on the question for which it is a coded answer. In other words, the coded portions of an extended-entry table require a code book before they can be understood. An example of a rather primitive code book is given in the upper half of figure 2.3. Code books occur in virtually every enterprise whose management has required the development of systematic descriptions of its financial, technical, production, marketing, or other activities. As we shall see later, the ability to use existing code books (for example, the chart of accounts) in the construction of extended-entry tables can be one of the advantages that this type of table has over the limited-entry form.

Decision-Table Nomenclature

Figures 2.5 and 2.6 merely abstract and generalize what we already know from our discussion of specific examples in the preceding sections:

1. All the questions to be asked or conditions to be tested are to be found in the condition stub.
2. All the types of action are to be found in the action stub.
3. All the combinations of answers (or condition test results) that are significant to the procedure are to be found in the condition entry; each such combination constitutes the condition portion of a rule.
4. All the courses of action permissible in the procedure are to be found in the action entry; each such course of action constitutes the action portion of a rule.
5. The correspondence between questions and answers, and between action descriptions and action specifications, is established by horizontal lines, which define *rows.*
6. The correspondence between sets of conditions and courses of action is established by vertical lines, which define *columns,* called *rules.*

Subprocedures, Procedures, Applications

In practice, most decision tables will not describe an entire procedure or system of procedures. They will describe either a portion of a main procedure or one procedure out of many in a "system" or "application."

Conventions are required: to tie together subprocedures into the system of which they are a part; to display authorship and time of preparation; to indicate which tables or programs become active after the subprocedure described by a table has been performed; and so on.

Figure 2.7 shows how a "functional" decision table might be embedded in a form that provides descriptive and connective information. The "Decision Table Body" portion contains the functional decision table. The "Exit Row" is also functional; exits are essentially specialized actions. The exit row can be thought of as an extension of the action section of the table; the action it describes is a branch to, or determination of, the next step to be taken. This can be either another table, a procedure described in some form other than a table, an error indication, or an indication that the procedure has been carried to completion.

The "Decision Table Header" is intended to be merely representative of some of the things that might be included as identification of a particular table. From the standpoint of this book, the only functional part of the header is the set of rule numbers. These provide numerical identification for the rules describing the decision-table procedure.

Incompletely Specified Rules

Figure 2.8 describes a real procedure—the one specified by the U.S. Internal Revenue Service for determining whether or not you are required to file a federal income tax return. The procedure is described by the IRS in two forms: (1) a narrative description and (2) a flowchart. In the next chapter, these are discussed and compared with each other and with figure 2.8.

At this point, the reader should be able to identify figure 2.8 as a limited-entry table and deduce that X as an action entry probably means "Perform the action," and that the absence of an X probably means "Don't perform the action." But in the condition entry there is something we have not yet discussed. Some of the questions are not answered. What does this mean?

The natural assumption, and the right one, is that, when answers have been omitted in the condition portion of a rule, they are not needed; the answers that *have* been given are sufficient to determine a course of action without the need for further questions. For example, the answer "No" to the first question in the condition stub of figure 2.8 is enough to determine a course of action: acquisition of publication 519. The answer "Yes," however, necessitates further questioning.

Nine questions are shown in the condition stub. However, only two rules, 12 and 13, require that all nine be answered. In all the others, one or more questions can be ignored. Ignoring a question is equivalent to making the answer "Don't care" and a blank condition entry square is commonly called a *Don't-care* entry.

Virtually all significant decision tables will have Don't-care entries. If they did not, decision tables would grow to unmanageable size. If all possible combinations of answers were significant for figure 2.8, the resultant table would have 512 rules. This would require a rather large volume of paper for what are, after all, nine rather simple questions. If the questions were to be as many as twenty, the resulting complete number of rules would be over a million.

Now one of the virtues we shall claim for decision tables is that they make it possible to check procedures for completeness and consistency. How is this possible when some questions go unanswered? The reader who wishes to anticipate for himself how tables with Don't-care entries can be checked for completeness is urged to try to determine whether or not figure 2.8 specifies completely what to do for any combination of answers to the questions in the condition stub. (*Hint:* It clearly would not be complete if one of its rules were omitted. How do we know that any possible additional rule is covered by one of the rules in figure 2.8?)

Vertical Decision Tables

To interpret the rules in figures 2.1 through 2.8, we have to read them from top to bottom rather than from left to right. This is because we have placed the condition and action stubs at the side of the table rather than at the top. In figure 2.9, the condition and action stubs are placed at the top and, as a consequence, the rules can be interpreted in the same order as the ordinary "if . . . then" statements to which they correspond. For many purposes, this form of table (particularly with some additional changes in organization) will be found more useful than the "standard" form illustrated in figures 2.1 through 2.8.

C. SUMMARY: DECISION TABLES IN CONTEXT

Decision Tables and Systems Analysis

Section 1B asserted the opinion that decision tables constitute a first attempt to develop a conceptual tool for the use of the systems analyst. Now that we have established a preliminary understanding of what decision tables *are*, it is possible to examine this opinion a little more closely.

For concreteness, let us consider a specific situation—that of the systems analyst who is asked to improve the efficiency of clerical operations in a tin mill.

In order to get his bearings, the first question he must ask is something like "What goes on here?" or "What do you do in the tin mill?" or "What *actions* are required to make tin plate?"

The answers he receives may or may not be intelligible to him. They might be, for example, "We pickle, clean, cold roll, anneal, temper, side trim, shear, and dip." Whether or not he understands these terms, he has a standard place in which to write them down: the *action stub*.

Having written down these comprehensible or incomprehensible terms, he can now ask the question, "How do you determine what specific things to do when you are asked to produce tin plate?"

The answers he gets now will largely contain the information he needs to fill out the *condition stub*. However, they will also frequently contain references to actions that were overlooked in the answer to his original question, "What do you do?" For example, the answer he gets might be "Well, it depends on who placed the order and what it's for. If it's one of the big companies like Intergalactic Can or Cosmic Can, then we probably schedule part of the order from some of our semifinished or finished inventories. However, if it's a small rolling for a little company like Universal Dog Food, we may schedule it right through from hot bands to coils or sheared plate. What we actually schedule depends on the product. Each product will require a different set of operations, and we will have to fit its production into schedules for each of the production units in the mill." The complete answer then extends into a discussion of specific products, what production units are required to make them, how those units are to be scheduled and controlled, and other details of manufacture.

Much of this new information will go into the condition stub. The systems analyst now knows that his condition stub must contain questions about—

1. The product.
2. The company placing the order.
3. The existence of semifinished inventories.
4. The existence of finished inventories.

However, he has also learned about some new entries for the action stub. In the original discussion, he was not told that the tin mill included not only *production* actions but also *inventory* actions—the actions required to maintain and keep track of semifinished and finished inventories.

At this point, he is probably prepared to retire, lick his wounds, and attempt to make sense of the information he has recorded in the *condition* and *action* stubs. He does this by expressing his understanding in terms of what we have called rules.

Inevitably, his initial understanding will be incomplete. He finds this out when he tries to write down rules and finds that he can't: he has failed to ask questions that he should have asked; he doesn't really understand the answers he received to the questions he did ask; some of the answers he got contradict others; some seem to be irrelevant; and so on.

What he should do now is prepare himself to ask more questions. He does this by preparing a decision table representing his current state of understanding, a decision table full of contradictions and irrelevancies and with incomplete rules—rules where he knows what conditions can exist but doesn't know what actions they imply, or where he knows what courses of action actually occur, but doesn't know under what conditions they are appropriate.

He should then go back and ask more questions, using the trouble spots in his decision table as a guide. When he does this, he will almost certainly learn of new actions, new conditions, and new rules. He should then modify his decision table accordingly and retest it for completeness, consistency, and relevance.

Thus, what the decision table can provide is an orderly mechanism for conducting a series of information-gathering interviews, each interview building in a systematic manner on the interviews that preceded it. Technical terms are isolated in two standard places—the stubs—where they provide a kind of glossary. Interconnections between condition sets and action sets are displayed in the entry portion of the table in a standardized manner that portrays relationships precisely, concisely, and vividly. The progress of a systems analyst's understanding can be measured by his ability to complete the entry portion of his table—that is, by the number of rules he can write to interconnect conditions and actions. His work is done when three conditions are met:

1. The *action stub* of his table contains descriptions of all the actions significant to the procedure he is describing.
2. The *condition stub* contains descriptions of all the factors to be considered in determining a course of action.
3. The *entry* portion contains a set of rules that are complete and consistent—that is, rules that prescribe one and only one course of action for each possible combination of conditions.

The Mechanics of Decision-Table Construction

It is unfortunate, but apparently inevitable, that books are better at illustrating end results than they are at illustrating the processes by which the end results were achieved. After we have completed a decision table, we know how many conditions, actions, and rules it comprises. Until we have completed it, however, we cannot be sure; we can only be sure after we have carried out the long, painstaking series of iterations that we discussed rather sketchily in the preceding section.

When we consider the completed decision tables that are used as illustrations in books about decision tables (this one, for example), we see that, while they are useful in illustrating ideas and describing small, completed procedures, they are impractical either for procedures in intermediate stages of development or for completed procedures that are very large.

Large procedures will have many rows and many columns. This means that an ideal decision table form should be capable of extending indefinitely upward to accommodate additional condition rows as they are needed, indefinitely downward to accommodate additional action rows, and indefinitely to the right to add new rules. This is clearly impractical.

It is even more impractical when we consider that in the course of developing the procedure we will be making tentative assumptions about all the decision-table components. As we progress, we will be changing these components—modifying, adding, deleting, redefining, and so on. Any fixed-format table we have prepared will have to be discarded as these activities change its basic structure. This will entail an enormous waste of paper, perspicuity, and patience.

In practice, then, we will usually find it impractical to use the fixed-format decision table for anything except, perhaps, documentation of the final procedure. Even here, the fixed-format table is apt to prove cumbersome or unwieldy because of its sheer physical size.

Can we salvage the important functional characteristics of a decision table by incorporating them in some kind of flexible format? What do we need that the fixed-format table cannot supply?

Briefly, what we need is something that might be called *indefinite extensibility*. If we want to discard a condition because it is either superfluous or requires redefinition, we'd like to be able to do it merely by crossing it out. If we wish to add either a replacement or a completely new condition we'd simply like to write it in.

Functionally, this might be achieved by a loose-leaf notebook divided into sections by ordinary index dividers. For relatively simple problems, three divisions might be adequate: Condition Stub; Action Stub; Rules. For more complicated problems, five sections might prove more useful: Condition Stub; Action Stub; Condition Sets; Action Sets; Rules. By *condition sets* and *action sets,* in the latter case, we mean combinations of conditions or actions that are actually significant to a procedure. As we have seen, in virtually no realistic case will all possible combinations occur. For complicated cases, it can be relatively difficult to determine just what combinations do occur, even without considering how a particular set of conditions is related to a particular set of actions.

If we assume that our basic flexible-format decision table takes the form of a loose-leaf notebook divided into five sections, then we have a relatively simple way to keep the first four sections up to date—that is, to have them reflect our current understanding of the procedure we are trying to describe. The indefinite extensibility of which we spoke earlier is that of the written page. We keep making entries down the page until we come to the bottom. Then we add a new page.

With the last section, Rules, we do the same as with the others except that we use the vertical form illustrated in figure 2.9. This permits us to add new rules in the same way we make new entries in the other four sections.

No matter what we do, complicated procedures will still be complicated. The best we can hope to achieve is some systematic way to cope with complexity. The functional sections of a flexible-format decision table can provide the basis for a practical way to master procedural complexity.

Decision Tables and "Information Systems"

When clerical operations are mechanized, the resulting agglomeration of people, machines, programs, cards, tapes, disks, forms, files, reports, and so on is frequently called a "management information system."

This usage is unfortunate. It suggests a neatness, consistency, and comprehensibility that are rarely attained by any sizable assemblage of real people using real machines to do real jobs. People come and go; machines and programs are continually being updated, expanded, or replaced; forms, files, reports, and information requirements are in a continual state of flux. As a result, instead of the order implied by the word *system,* we commonly find a disorder so great that, to the uninitiated, it is apt to be seen as a kind of controlled chaos. It is a sad fact that information systems are themselves very rarely systematic.

This is true whether or not the "information system" is mechanized. The only difference mechanization makes is to force some kind of solution to problems that existed, but were ignored, in unmechanized systems. This throwing open of the closet doors behind which skeletons have long been concealed is frequently one of the unintended, but very important, benefits of mechanizing clerical operations.

The same kinds of problems arise no matter which portion of an information system we study. The "software" required to operate a modern computer installation, for example, includes operating systems; language translators; loaders; report generators; sort/merge programs; packaged applications; debugging aids; utilities; subsystems to perform such specialized tasks as paging, spooling, scheduling, and logging; "access methods" to control tapes, typewriters, drums, card readers, disks, card punches, plotters, cathode-ray tubes, printers using a wide variety of character sets, sensing and monitoring instruments, other computers ... A complete list of the subsystems covered by the term *software* would be long indeed.

Though this sounds complicated, the software system is by far the best documented and best understood part of the information system proper—with the exception of the physical devices themselves. Software systems are supported by reference manuals, program logic manuals, flowcharts, courses of instruction, self-instructional material, system engineers, and a variety of other aids.

The very bulk, complexity and cost of the support for a software system suggests the need for a better way to provide this support. This need grows as the systems themselves proliferate and grow. But at least the problem is recognized and, however inadequately, solved.

This is not true of the system of primary interest to us—the management information system itself. In general, we do not understand it and cannot describe it as well as we can its software component. This is not for want of trying. Large corporations usually have whole departments full of systems analysts, procedure designers, industrial engineers, technical writers, and other specialists whose mission it is to understand, analyze, design, and describe management information systems. Most of them do the best job possible with the tools available. But the tools are inadequate to the immensity of the task. Instead of unified, integrated descriptions of entire systems, most organizations have to make do with sketchy, variegated descriptions of the loosely coordinated subsystems that make up the whole.

Neither decision tables nor any other technique currently available can solve these problems easily or quickly. But decision tables, if employed rigorously and imaginatively, give us reason to hope that they can be made more tractable. This hope will be realized if the properties of decision tables are effectively exploited. Some of the more important of these properties are described in the remaining chapters of this book, each of which consists of a relatively brief treatment designed to suggest ideas and stimulate thought rather than provide final answers or definitive pronouncements.

REVIEW

Decision tables are structured procedure descriptions. In form, they are rectangles divided into four sections by double lines. The upper left section is called the *condition stub*. It contains a set of explicit or implicit questions covering the entire range of *conditions* that must be considered before selecting a course of action. The lower left section is called the *action stub*. It contains the entire set of individual *actions* governed by the decision table. A *course of action* is a specification of a set of these individual actions.

The *upper right* section is called the *condition entry*. It consists of answers to the questions on the condition stub, the answers along any *row* corresponding to the question on that row in the condition stub. (Rows are delineated by horizontal lines.)

The *lower right* section is called the *action entry*. It consists of specifications of which of the actions described in the action stub is to be performed, the specification along any row corresponding to the action described in that row in the action stub.

The association between specific combinations of answers and specific combinations of actions is established by *rules*, represented by the *columns* into which the right-hand (or *entry*) section of the decision table is subdivided. (Columns are delineated by vertical lines.) The general interpretation of any rule is as follows:

If the questions asked in the condition stub are answered in the way specified by the condition portion of the rule, then the actions to be performed are those prescribed in the action portion of the rule.

Two specific interpretations will illustrate the general idea:

Rule 4, figure 2.1

In words, this is the rule: If you sell a Reo that's running well, your commission is 5 percent. You need not schedule any shopwork and you need not get a manager's O.K. for the sale.

Rule 10, figure 2.8

If you are a U.S. citizen or resident *and* your gross income is greater than $600 but not greater than your minimum taxable gross income *and* you are married *and* no other person can claim you or your spouse as an income tax exemption *and* you are filing separate returns then you must file a return.

In general, the individual procedure described by a decision table is one of many individual procedures describing an *application*. Thus each table requires, in addition to its *functional elements* (condition stub, action stub, condition entry, action entry), the addition of information about authorship, date of preparation, the application or higher-level procedure of which the decision table is a part, the next procedure to be followed, and so on. Figure 2.7 shows a representative form for displaying this kind of ancillary information.

The two broad categories of decision tables are distinguished by the kinds of questions they ask and the way they specify which actions to take.

The *limited-entry* table asks yes-no questions. It specifies actions in a yes-no manner also—that is, by listing all possible actions and then specifying either "do" or "do not" for each of them.

The *extended-entry* table asks more general questions and characterizes actions more broadly. Each question is about a significant *attribute* (for example, *make* or *condition*); each action is specified as an action *type* (for example, *commission, shopwork, manager approval*). In this kind of table, an answer consists of specifying a value for an attribute like *make* (Reo) or *commission* (10 percent).

Either words or codes can be used in the condition and action entries of a decision table. In the latter case, the table is said to be in *coded form*.

In practice, most tables will contain rules in which some of the questions in the condition stub are ignored. When a square in the condition entry is blank, it is called a *Don't-care* square, implying that the course of action prescribed by the rule would be the same whatever answer was placed in the square.

EXERCISES

In the first four questions, portions of decision tables are displayed. Fill in the blanks below each figure by naming the part of the table that the figure illustrates and the type of table into which it would fit.

.25	.30	.17	.12	.14	.39	.41
17	12	14	19	21	11	30
Sulfur	Manganese	Copper	Copper	Aluminum	None	None

1. This could be a (n) _____ of
   a (n) _____ -entry table.

   Is the book a paperback?

   Is it a used book?

   Does it have a reorder slip?

2. This could be a (n) _____ of
   a (n) _____ -entry table.

   Pct. scrap in total charge

   Heat time allowance (minutes)

   Major additive

3. This could be a (n) _____ of
   a (n) _____ -entry table.

A
2
Q
Z
14
R

4. This could be a (n)_____ of a (n)_____-entry table.

5. Write out your own description of the Cord-Reo-Duesenberg procedure, using only figures 2.1 to 2.4 as a guide.

6. Do the same for the Filing Returns table of figure 2.8.

7. If you are familiar with flowcharting, draw flowcharts for the two procedures you described in answer to exercises 5 and 6.

8. The following is a verbal description of the procedure defined by a flowchart at the beginning of chapter 16, "Your Federal Income Tax":

   Definition:
   "Sick Pay"—Payment for absence from work because of sickness or injury, not necessarily in connection with your work.

   Instructions:
   a) If you received sick pay at a rate of 75 percent or less of your regular pay for at least one day of absence during which you were hospitalized, read chapter 16 to determine if you qualify to exclude sick pay from your income.
   b) If your absence exceeded seven days and your rate of sick pay was 75 percent or less of your regular pay, read chapter 16.
   c) If your absence exceeded 30 days, read chapter 16.
   d) If none of these conditions hold, you do not qualify for sick pay exclusion and need not read chapter 16.

   Construct a decision table for this procedure.

9. Decision table "actions" can cover a broad variety of forms, some of them almost purely descriptive. Consider a decision table that might be useful to a computer dating bureau that has been asked to find com-

patible female partners for the members of a baseball team. Its survey develops the following preferences by team members:

- a) Outfielders prefer redheads. Other team members do not.
- b) Left-handers prefer girls with glasses. Other team members prefer girls without glasses.
- c) Rookies feel happiest in the company of women over forty. Others prefer less mature companionship.

(1) Prepare a decision table that will express these preferences. (2) What kind of girl would please a right-handed pitcher who has been playing in the major leagues for twenty years?

10. The varied difficulties that confront the systems analyst derive primarily from one fundamental problem, combinatorial complexity—the surprisingly large number of combinations he finds himself forced to consider when he attempts to understand even relatively simple procedures. It is this complexity that provides the challenge and entertainment in puzzles like the ones that start out "The Ukrainian drinks slivovitz," "The owner of the pink house owns a poodle and chews Bull Durham,"..., and end with the question "Who owns the camel?"

    A trivial puzzle of this kind is offered below. Can you use any of the ideas underlying decision tables to describe and solve this puzzle in an orderly manner? (Disregard the fact that in this case it is much easier to solve it in a disorderly manner.)

    - a) The pet of the man in the blue house was once nominated for "Fox of the Year" by the FFV.
    - b) The man in the red house has an inordinate fear of otters and will not visit the neighbor who keeps one as a pet.
    - c) Tom lives next to the white house.
    - d) Tom and Harry feel that foxes are a disgrace to the neighborhood and have threatened to bring suit against Dick for keeping one.
    - e) Harry's sister married the man who owns the alligator.
    - f) The puzzle involves only three houses, three owners, and three pets.
    - g) Who owns the alligator?

11. If we examine illustrations 2.2 and 2.4, we see that some of the answers in a rule imply or *force* answers to other questions in the same rule. This is because the questions are not independent of one another; we cannot meaningfully describe one car as being both a Cord and Reo, for example (our hypothetical dealer does not deal in hybrid cars). Neither can we specify a commission as both 5 percent and 10 percent in the same rule. Examine figure 2.2 and determine which rows are dependent and which independent.

	1	2	3	4	5	6	7	8	9
MAKE	Cord	Cord	Cord	Reo	Reo	Reo	Duesenberg	Duesenberg	Duesenberg
CONDITION	Running well	Running poorly	Not running	Running well	Running poorly	Not running	Running well	Running poorly	Not running
COMMISSION	5%	10%	10%	5%	10%	10%	Variable	Variable	Variable
SHOPWORK	Not required	One week	Six weeks	Not required	Two weeks	Six weeks	Estimate	Estimate	Estimate
MANAGER O.K.	Not required	Not required	Not required	Not required	Not required	Not required	Required	Required	Required

FIGURE 2.1   Vintage-car decision table: Extended entry

	1	2	3	4	5	6	7	8	9
Is car a Cord?	Yes	Yes	Yes	No	No	No	No	No	No
Is car a Reo?	No	No	No	Yes	Yes	Yes	No	No	No
Is car a Duesenberg?	No	No	No	No	No	No	Yes	Yes	Yes
Is car running well?	Yes	No	No	Yes	No	No	Yes	No	No
Is car running poorly?	No	Yes	No	No	Yes	No	No	Yes	No
Is car not running?	No	No	Yes	No	No	Yes	No	No	Yes
Commission is 5%	Yes	No	No	Yes	No	No	No	No	No
Commission is 10%	No	Yes	Yes	No	Yes	Yes	No	No	No
Commission is variable	No	No	No	No	No	No	Yes	Yes	Yes
No shopwork needed	Yes	No	No	Yes	No	No	No	No	No
Schedule 1 week of shopwork	No	Yes	No	No	No	No	No	No	No
Schedule 2 weeks of shopwork	No	No	No	No	Yes	No	No	No	No
Schedule 6 weeks of shopwork	No	No	Yes	No	No	Yes	No	No	No
Estimate shopwork	No	No	No	No	No	No	Yes	Yes	Yes
Get manager O.K.	No	No	No	No	No	No	Yes	Yes	Yes

FIGURE 2.2   Vintage-car decision table: Limited entry

CODES FOR DECISION TABLE OF FIGURE 2.1

MAKE	CONDITION
1. Cord	1. Running well
2. Reo	2. Running poorly
3. Duesenberg	3. Not running

COMMISSION	SHOPWORK	MANAGER O.K.
1. 5%	1. Not needed	1. Not required
2. 10%	2. One week	2. Required
3. Variable	3. Two weeks	
	4. Six weeks	
	5. Estimate	

MAKE	1	1	1	2	2	2	3	3	3
CONDITION	1	2	3	1	2	3	1	2	3
COMMISSION	1	2	2	1	2	2	3	3	3
SHOPWORK	1	2	4	1	3	4	5	5	5
MANAGER O.K.	1	1	1	1	1	1	2	2	2

FIGURE 2.3 Vintage-car decision table: Coded, extended entry

	1	2	3	4	5	6	7	8	9
Is car a Cord?	1	1	1	0	0	0	0	0	0
Is car a Reo?	0	0	0	1	1	1	0	0	0
Is car a Duesenberg?	0	0	0	0	0	0	1	1	1
Is car running well?	1	0	0	1	0	0	1	0	0
Is car running poorly?	0	1	0	0	1	0	0	1	0
Is car not running?	0	0	1	0	0	1	0	0	1
Commission is 5%	1	0	0	1	0	0	0	0	0
Commission is 10%	0	1	1	0	1	1	0	0	0
Commission is variable	0	0	0	0	0	0	1	1	1
No shopwork needed	1	0	0	1	0	0	0	0	0
Schedule 1 week of shopwork	0	1	0	0	0	0	0	0	0
Schedule 2 weeks of shopwork	0	0	0	0	1	0	0	0	0
Schedule 6 weeks of shopwork	0	0	1	0	0	1	0	0	0
Estimate shopwork	0	0	0	0	0	0	1	1	1
Get manager O.K.	0	0	0	0	0	0	1	1	1

FIGURE 2.4   Vintage-car decision table: Coded, limited entry

Condition stub	Condition entry
Action stub	Action entry

FIGURE 2.5 Decision-table nomenclature:
    I. Stubs, entries, conditions, actions

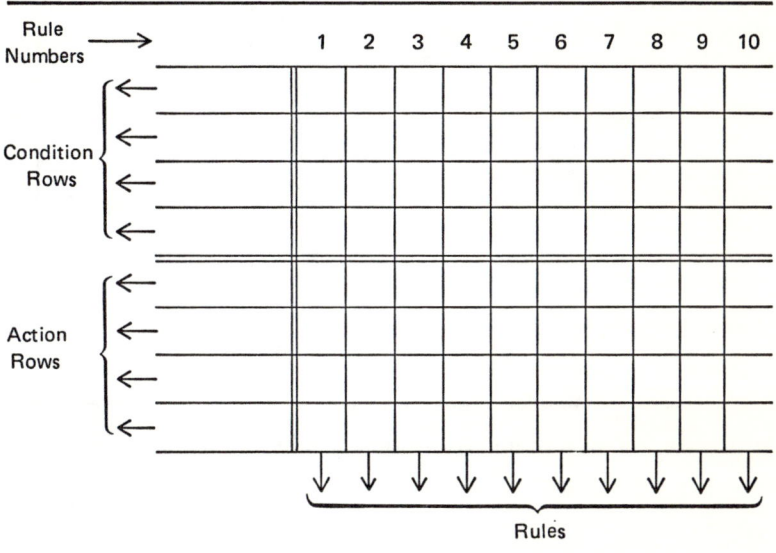

FIGURE 2.6 Decision-table nomenclature:
    II. Condition rows, action rows, rule columns

27

Header	Application Name:	Prepared by:													
	Program Name:	Date:													
	Table Name:	1	2	3	4	5	6	7	8	9	10	11	12	13	14
Body															
Exit Row	Exit to														

FIGURE 2.7 Decision-table nomenclature: III. Ancillary information

Codes for Decision Tables 0: No 1: Yes	1	2	3	4	5	6	7	8	9	10	11	12	13	14
Are you a U.S. citizen or resident?	0	1	1	1	1	1	1	1	1	1	1	1	1	1
Is your gross income greater than $600?		0	0	0	1	1	1	1	1	1	1	1	1	1
Is your gross income greater than your minimum taxable gross income? *					1	0	0	0	0	0	0	0	0	0
Were you married as of the end of the year?						0	0	0	1	1	1	1	1	1
Can some other person claim either you or your spouse as an income tax exemption?									1	0	0	0	0	0
Are you filing separate returns?											1	0	0	0	0
Were you both living in the same household at the end of last year?											0	1	1	1
Did you have either self-employed income over $400 or tips on which social security was not reported?		0	0	1		0	0	1			0	0	1
Was any income tax withheld from your pay?		0	1			0	1				0	1	
Obtain publication 519.	x													
File a tax return.			x	x			x	x	x	x			x
Do not file a tax return.		x				x						x		
File for a refund.			x				x						x	

*Minimum Taxable Gross Income Table

Single, under 65	1700
Single, 65 or over	2300
Married, both under 65	2300
Married, one under 65	2900
Married, neither under 65	3500

FIGURE 2.8   Income tax decision table: Filing returns

MAKE	CONDITION	COMMISSION	SHOP WORK	MANAGER O.K.
Cord	Running well	5%	Not required	Not required
Cord	Running poorly	10%	One week	Not required
Cord	Not running	10%	Six weeks	Not required
Reo	Running well	5%	Not required	Not required
Reo	Running poorly	10%	Two weeks	Not required
Reo	Not running	10%	Six weeks	Not required
Duesenberg	Running well	Variable	Estimate	Required
Duesenberg	Running poorly	Variable	Estimate	Required
Duesenberg	Not running	Variable	Estimate	Required

FIGURE 2.9   Vertical form of Figure 2.1

3

Narrative Descriptions and Flowcharts

Offhand, everybody knows what English is. It is the language we have been using all our lives, it is reading and writing, it is what we all studied for years in school—grammar and spelling, composition and literature. Everybody seems to agree too that it is the most fundamental subject, inasmuch as it is the only one required of all students throughout their school years; all are obviously dependent on their command of the language for both their learning and their living purposes. But here is the beginning of much trouble. Nobody is likely to think very hard about something that everybody takes for granted. Until recent years educators gave remarkably little attention to this most fundamental subject in the curriculum, even though it was plain that after all those years in school most young people could not read, write or speak well. For such reasons, the Anglo-American Seminar opened and ended its discussions with the seemingly elementary question: What is English?

> Herbert J. Muller, *The Uses of English: Guidelines for the Teaching of English from the Anglo-American Conference at Dartmouth College* (Holt, Rinehart & Winston, 1967). Reprinted by permission.

A. AN UNNATURAL USE FOR NATURAL LANGUAGE

It is customary in the literature of data processing to speak of automatic coding schemes like FORTRAN, COBOL, BASIC, and ALGOL as "programming languages." As a result, we have had to find a new way to describe English, Italian, Chinese, and Russian. We used to call them simply "languages"; in the context of computers they are now referred to as "natural languages."

The implication is clear. If only manufacturers would produce computers that could be programmed in a "natural" language, professional programmers would no longer be needed; anyone who could write could program. But how many people can write? How many good programmers do you know? How many good writers? Which is the rarer skill?

Clear writing, like clear thinking, is hard work. It is *not* "natural," if by "natural" we mean *spontaneous* or *automatic* or *effortless*. A reader fortunate enough to find something easy to read can be sure that the writer found it hard to write.

Why should this be? We have all been using and studying a "natural language" since infancy. Why should it be so hard to use natural language to say what we mean?

Any comprehensive discussion of this question is clearly beyond the scope of this book and certainly beyond the competence of its writer. But the concluding question of the preceding paragraph suggests its own answer by making two fallacious assumptions which require rebuttal:

1. that natural language is natural
2. that saying is the same as writing

Neither of these assumptions is true.

It is clear that no language is truly natural. If it were, then all the world would be of one tongue and none of us would have needed the long years of instruction in English that we received in grammar school, high school, and college. Acquiring a language would take as little conscious effort as breathing, or perspiring, or letting the hair grow. Instead the painful reality is that after all our years of instruction we still find it hard to say precisely what we mean in plain English.

Important as it is to realize that no language is natural, for our purposes the second part of the rebuttal is even more important. This whole book is really about a more effective way to write procedure descriptions. Therefore, it is of paramount importance to remember that *writing is not saying.*

When we use the word *say,* we suggest the one use of language that comes closest to being natural: oral, face-to-face communication between two people—*dialogue.* The most important characteristic of dialogue is that it is two-way communication. If a participant in a dialogue says something that is not clear to his auditor, he can be stopped by a question, a puzzled look, or a snore. The point of difficulty can then be discussed until it is either eliminated or recognized as an irreconcilable difference in outlook, interest, or comprehension. It is this continual give-and-take that makes dialogue—imperfect as it usually is—the most effective form of communication between people.

A writer—particularly a writer of procedure descriptions or technical reports or textbooks—does not participate in a dialogue. His communication is all one way. As he labors over his monologue, he must do his best to guess what he should explain and what he can assume the reader will know without explanation. Since he has no way of knowing who his reader is—or even if he will have one—his guesses can be quite wide of the mark. Yet guess he must if his report or book is ever to get written. In essence, what the writer must guess is which of the words he puts down will have the same meaning for his reader as they have for him, and which will either have no meaning at all or will be understood to have some meaning other than what he intends. For example, in chapter 2, the word *pickle* occurs in a discussion of tin mill operations. What does it mean to the reader? What does it mean to a tin mill worker? When is the writer required to define a term and when can he use it without definition?

The point is that isolated words do not have definite meanings; most of the meaning they convey depends on the words that surround them and on when, where, how, why, about what, to whom, and by whom they are used. They change from one time to another, from one place to another, from one culture to another, and—to express all the contrasts generally—from one context to another. In Shakespeare's time, the word *pitiful* meant "filled with pity"; in our time, it has the inverse meaning "arousing pity." To the layman, *pickling* means soaking a cucumber in brine; in the tin mill, it means passing a "hot band" through a solution of sulfuric acid. Further, when two words whose individual meanings are relatively definite are combined into a compound (to provide a name for a new thing or idea), the meaning of the individual words usually tells us very little about the meaning of the compound. *Data* and *processing* tell us very little about the vast assemblage of machines, ideas, and activities that we now call data processing. *Decision* and *table* tell us very little about the subject matter of this book.

The fallacy that "natural language" is the best way to "communicate with a computer" (in itself an expression so absurd that it would provoke laughter if it did not provoke tears) permeates most of modern data processing. It has led to the development of programming languages that use symbols similar in appearance to the words of everyday speech. In all cases, the most obscure, most unnatural parts of a programming language are those that use the symbols of a natural language. "READ" and "WRITE" as they occur in COBOL, FORTRAN, and BASIC, have very little in common either with the English words they resemble or with each other. Indeed, far from being good for "communicating with a computer," natural language is not even good for communicating with people, if the communication is about the kind of precise, intricate, specialized topics that occur in science, law, systems analysis, programming, or any other endeavor requiring an exact communication of meaning.

Since our primary concern here is with systems analysis, we can most profitably examine this point by studying a good procedure description and seeing some of the ways in which it departs from the informal, day-to-day use of the language in which it is written. This is what we shall do in the remainder of this section. However, the reader who is unconvinced that "plain English" is far from simple is urged to consider carefully the first five exercises at the end of this chapter. They are intended primarily to stimulate him to start—and continue—his own investigation of how natural language is really used to communicate and where natural language is and is not effective. The subject is well worth his careful attention.

Figure 3.1 provides an example of a good procedure description.* It is the verbal equivalent of the decision table shown in figure 2.8. It is written in English. How does the English compare with the English of literature, of personal correspondence, and of everyday speech?

Some major points of difference stand out at once:

1. The use of tables.
2. The use of itemized lists.
3. The large proportion of the text devoted to defining special categories of taxpayer or potential taxpayer.
4. The reference to other sources for more detailed information.
5. The large proportion of interdependent conditional statements, usually identifiable by words like *if* or *however*.

This is not everyday language. It is the language of a skilled technical writer attempting to convey a specific body of information precisely, concisely, and clearly. He has done as well as could be done with words. What can we say about the result?

Even about a procedure as relatively simple as this, anyone who really has any doubts about whether he is required to file a return might well experience some uncertainties:

1. Has he considered and understood all the defined categories which might apply to his situation?
2. Has he correctly traced all the chains of reasoning represented in the text by the scatter of *ifs, howevers,* and other condition-describing words?

---

*Figure 3.1 is copied directly from "Your Federal Income Tax," a yearly publication of the U.S. Internal Revenue Service. It is customary to berate the bureaucracy for the incomprehensibility of its publications. It is not as customary to praise it for a job well done. This publication is a job extremely well done. Anyone interested in learning what goes into a good procedure description might well use it as a model.

The first question is about what certain key words mean when they are used in the procedure; the second question is about how these key words are related to each other. The first is about building blocks; the second about how the blocks are put together.

All that we can say about the first question—the need for precise definition of key words—is that it is of fundamental importance no matter what form of procedure description is used. It not only requires care and skill in the use of words and other means of definition; it also requires good judgment and an accurate understanding of the audience for whom the procedure description is intended. In figure 3.1, "age 65" is defined. Why? Surely we all know what we mean by "age 65." If we examine the procedure, however, we see that the definition of "age 65"—for the purposes of this procedure—is different from the common understanding. The procedure writer knew this and so made a point of stressing the definition. On the other hand, two terms are *not* defined: "U.S. resident" and "bona fide resident of Puerto Rico." Why not? At what moment does a visitor to the United States turn into a resident? Is there such a category as *bogus* resident of Puerto Rico? Why aren't we told? One can only guess. The author's guess is that those who are legally defined as "residents" know who they are; those who are not need not worry about it. Again, the decision not to define is an exercise of judgment on the part of the procedure writer.

The second question is the one of primary concern to us. It is the one that motivates both the drawing of flowcharts and the construction of decision tables.

B. FLOW CHARTS: THE SERIAL DISPLAY OF BRANCHING STRUCTURE

Figure 3.2 is a flowchart copied from "Your Federal Income Tax"; it is the one referred to at the end of the narrative description of figure 3.1. It does not use the standard components (rectangles, circles, triangles, lozenges, schematic card, tape, printer symbols, etc.) used in most program flowcharts. Figure 3.3 is a flowchart for the same procedure. It uses three of the standard symbols: lozenges to identify decision points or "branches"; rectangles to identify actions; circles to identify connectors.

The use of lozenges and lines points up the flowchart's most important characteristics. The lozenges identify the places where questions are asked (or conditions are tested); the lines connect the various portions of the flowchart and make it easy to trace a path from one question to another (through intervening action boxes, if necessary). In comparison with the narrative description, it is much easier to separate conditions from actions and much easier to trace chains or sequences of consecutive conditions and actions. These are important aids to understanding and account in large part for the fact that flowcharts are, except for narrative descriptions, the most widely used technique for documenting procedures.

Other flowchart symbols to denote specialized actions are also in common use. They serve the same function for the processes they denote as the lozenges do for condition testing: to make it easy to scan a procedure description and locate the points at which specific types of action take place.

Why didn't the Internal Revenue Service use the standard flowchart symbols in the flowchart shown in figure 3.2? The answer is undoubtedly the practical kind of consideration that baffles so many schemes that are theoretically impeccable. It is hard to write long questions inside little diamond-shaped boxes. One has to abbreviate (in effect, code), or put the question outside the boxes, or make the boxes bigger, or resort to some other expedient. Each of these alternatives makes it harder to draw the flowchart and detracts from its comprehensibility, either by making it necessary to remember or translate codes, or by complicating the placement of boxes on the page, or by increasing the size of the flowchart, thus making it harder to connect one part with another. As can be seen from figure 3.2, the points at which questions are asked are, for this procedure, as easy to identify without specially shaped question boxes as they would be if all the questions had been identified by lozenges. The questions themselves are a good deal more comprehensible than they would be if they had been restricted to a length that would fit inside the rather inconveniently shaped standard symbols. For the purposes of a procedure description intended for the layman, use of the standard convention would have been at best a distraction and at worst a source of confusion.

Can we say anything about how narrative descriptions and flowcharts compare with decision tables?

C. NARRATIVE DESCRIPTIONS, FLOWCHARTS, AND DECISION TABLES: A COMPARISON

For our purposes, the primary advantages of the flowchart over the narrative description are the two we discussed in the preceding section:

1. The questions are easier to distinguish from the other components of the procedure.
2. Chains or sequences of questions and answers are easier to identify and trace.

What about decision tables? In comparison with the other two, we can say that in decision tables—

1. The individual questions are easiest of all to locate and understand. They are *all* in the condition stub so we can even tell how many *different* questions there are, something hard to determine in the other methods.
2. The same is true of individual actions, which are *all* defined in the action stub.

3. Most important of all, the decision table displays explicitly and clearly all significant *combinations* of conditions and actions, as well as the association between them. These do not appear explicitly in either the narrative description or the flowchart.

This last point deserves elaboration. To make it clear, let us examine figure 3.3 carefully and contrast it with figure 2.8, the decision table from which it is derived.

Figure 3.3 has some annotations which are *not* part of any standard flowchart conventions. They have been added to the flowchart to facilitate the discussion which follows. The additions consist of (1) the Roman numerals I-XIV, which appear next to the exit connectors (the connectors pointed to by an arrow) and (2) the decimal numbers that appear in parentheses below or to the left of the flow lines leaving a decision lozenge. (The latter will be used and discussed later in the book.)

The Roman numerals identify the end of a chain of decisions that determine a course of action. They serve the same function as the rule numbers of figure 2.8, though they are not numbered in the same order.

Let us trace the chain of decisions that culminates in the connector labeled XIII and compare it with the rule to which it corresponds, which in this case happens to have the corresponding Arabic numeral 13.

Once we have traced through the flowchart to get to connector XIII, we know where we are but are apt to forget exactly how we got there. In other words, we have no record of which set of branches we took in tracing our way through the flowchart; we do not even have a record of how many decision boxes we traversed. If we wish to check to make sure that we answered all the questions correctly, we must repeat the process. If we end up at the same point, our confidence that we are right is increased but an element of doubt could still linger, particularly in a more realistic, more complicated flowchart.

In the decision table, on the other hand, once we have located a rule (by a process similar to that of tracing through a flowchart, as the reader should verify for himself), we have definite knowledge of how *all* the significant questions must be answered if that rule is to apply. These answers are the condition set shown in the condition portion of the rule. In rule 13, for example, it is easy to check the fact that all nine questions were asked, that questions 1, 2, 4, 7, 9 were answered yes, and that all the others were answered no. In other words, once you have selected the rule that you believe covers your situation, you have an independent way of checking whether your belief is correct. You do not, as in the flowchart, have to repeat the selection process and hope you end up at the same place.

This advantage of the decision table is of fundamental importance. At present, it is insufficiently exploited, even by the most dedicated users of decision tables. What the decision table provides is a way to translate a *condition set*—a set of answers to questions—into a set of numbers. We can do the same for an action set—a set of action specifications. We can now describe our procedure by matching up condition sets and action sets. *Having done this, we have a form of procedure description that is ideal for automatic processing by digital computer.*

The implications for systematic, automatic analysis, design, and documentation are great. So are the implications for debugging (detecting and eliminating errors), maintenance, and redesign. Once a procedure has been defined as a specified correspondence between number sets, we can use a computer to do automatically, quickly, accurately, and inexpensively what is now done by hand slowly, inaccurately, and at great cost.

To suggest some of what this digital form of procedure description makes possible, let us discuss briefly the improvements in systems analysis that are possible now and the further improvements that can be achieved with a relatively modest expenditure of effort.

If we process a decision table by computer, we now can—

1. Check to see that we have considered all possible combinations of the factors that determine courses of action. (*Completeness.*)
2. Check to see that we do not contradict ourselves. (*Consistency.*)
3. Weed repetitive or unnecessary instructions out of a procedure. (*Elimination of redundancy.*)
4. Transform the complete, consistent decision table into either an efficient program or a flowchart to guide efficient, detailed programming. (*Program specification.*)

Let us defer until later in the book a discussion of how this can be done. Assuming that it can be, couldn't we do the same with either a conventional, narrative procedure description or a good flowchart?

The answer is no in both cases. There is no practical way to use a computer to convert written descriptions or flowcharts into efficient computer programs, to check them for completeness and consistency, or to eliminate unnecessary tests or instructions. Any way we might devise to achieve these objectives would almost certainly require us to impose a numerical (or, more precisely, a digital) superstructure on either the written description or the flowchart. This superstructure would, in its essentials, be equivalent to a decision table.

The advantages we have listed are important ones that anyone concerned with the efficiency of his data-processing system can achieve now. Advantages which could very well be even more important can be achieved with a relatively modest development effort.

D. SYSTEMS ANALYSIS: MECHANIZING THE PAPER-SHUFFLING

Good systems analysis and good programming alternate between creativity and drudgery. In designing a procedure, a good systems analyst or programmer gets his satisfaction from the relatively brief periods of creativity that make his job worthwhile; he earns this satisfaction by putting up with the long hours of drudgery that make it possible.

The drudgery is largely a matter of paper-shuffling. The facts and relationships that define a procedure must be recorded, cross-indexed, checked, and organized into a clear procedure description. Inevitably, an initial description will be modified time and time again. Each modification will necessitate a laborious rerecording, recross-indexing, rechecking, and reorganizing. The clerical burden is staggering. It is the biggest single contributor to the time and expense required to design procedures. It is the biggest single obstacle impeding the design of large, unified, comprehensive procedures which might truly merit the honorific "integrated data-processing system."

Computer assistance in eliminating the drudgery of these tasks is now at hand. The promise of time-sharing has been realized. Reliable, relatively inexpensive time-sharing systems are now available to many people and there is every reason to believe that their use will become even more common as their costs decrease and their capabilities increase.

What this means is that it is now possible to eliminate a great deal of the drudgery of systems analysis by providing the systems analyst with a typewriter connected to a computer. All it takes is the effort required to develop a program that will do the following:

1. Accept and code information that defines condition stubs and action stubs.
2. Accept and code information that describes rules.
3. Apply completeness, consistency, and redundancy checks as needed.
4. Print out tables, subtables, or individual rules on demand, either on the line printer for documentation or on a typewriter or cathode-ray tube for examination by the systems analyst.
5. Print out a code book and flowchart to provide specifications for programming.

Further discussion of how this can be achieved is clearly beyond the scope of an introductory book. But the chapters that follow describe the mechanics of checking a coded decision table for completeness, consistency, and redundancy, and indicate how such a table can be converted into a flowchart or program. These methods can be made the basis for programs with more ambitious objectives.

E. SUMMARY: PROCEDURES, ALGORITHMS, PROGRAMS

Three terms are commonly employed in discussing data-processing systems: *procedure, algorithm, program*. These are different names for the same thing: a sequence of elementary actions that will produce a desired end result. An invoicing procedure is a sequence of actions that will produce a bill. A square root algorithm is a sequence of actions that will produce a square root. Either of these may be written as a program that will, in the first case, produce a bill and, in the second, a square root. They are *all* procedures; the only difference is that the last two terms are usually restricted to specific areas of application. *Algorithm* usually means a mathematical procedure. *Program* usually means a computer procedure.

What we have done in the preceding section of this chapter is eliminate the distinction between what are ordinarily called procedures—invoicing, payroll, accounts receivable, and so on—and the specialized procedures called algorithms or programs.

We have done this by highlighting the distinction between the two different parts of a procedure that are ordinarily intermixed:

1. Definitions
2. The description of structure

Definitions tell us what a procedure *means*. They describe the building blocks of a procedure in terms that relate a procedure to the world external to the procedure, the world in which the procedure serves a specific function: producing a bill, a square root, a pay check, an exponential integral, a cost report, and so on.

The description of structure tells us what a procedure *is*. It describes the world internal to the procedure, independent of the meaning of the building blocks. The structure of a simple invoicing subroutine, for example, might well be the same as the structure of a simple subroutine to calculate gross pay. In the invoicing subroutine, our input might be two sets of numbers: the first set identifies the items that have been sold; the second set tells how much of each item has been sold. The structure of the procedure then consists of three operations:

1. *Table lookup,* in which the item numbers are used to search a table that tells us the price of each item.
2. *Multiplication* of the prices by the number of items sold.
3. *Addition* of the multiplied quantities to get the total amount of the bill.

In the gross pay subroutine, our input is two sets of numbers which are treated in exactly the same way, though they are defined differently. The first set identifies either types of work or types of manufactured article; the second specifies either numbers of hours or numbers of pieces. The processing is the same as for the invoice:

1. *Table lookup* to determine the hourly rate or the piece rate.
2. *Multiplication* by number of hours or number of pieces.
3. *Addition* to calculate gross pay.

The distinction between definitions and structure is important, because this book is concerned primarily with structure. We are not concerned with what procedures do but with how they do it. That is, we are concerned not with specific procedures—invoicing, payroll, linear programming, and so on—but with procedures in general.

Our primary interest is in what we have called *program logic* or *branching structure*. In a coded decision table, program logic is described primarily in the entry portion of the table, the part to the right of the vertical double lines. Most of the following discussion will be devoted to a study of the condition entry section of coded, extended-entry decision tables.

REVIEW

A written procedure description is sometimes called a *narrative* description. When it is prepared by a skilled procedure writer, it provides an excellent description of the details of a procedure. It is less satisfactory when we wish to consider the procedure as a whole—that is, when we wish to consider its "flow" or structure.

A flowchart, as its name suggests, is a procedure description that uses lines and specially shaped graphic symbols to picture the flow of a procedure. A flowchart makes it possible to trace a sequence of conditions and actions more readily and more comprehensibly than is possible with most written procedure descriptions. Its structure is such that the places where questions are asked or conditions are tested are easier to locate quickly than they would be in a corresponding narrative description. The use of special symbols other than those identifying decision or "branch" points (for example, for tape, card, or printer operations) serves an analogous function: making it easier to locate the places in the procedure where these operations take place.

The flowchart is still unsatisfactory when we wish to picture the procedure as a whole. While it is easy to trace any particular path through a flowchart, it is difficult to get a picture of *all* possible paths. The decision table provides such a picture—the set of rules that appear in the entry portion of the table. These rules give us a compact picture of the procedure as a whole. In coded form, they constitute a *digital* description of a procedure. This kind of description not only is a more useful picture of the complete procedure for human use, it also makes it possible to use a digital computer to check the procedure for consistency and completeness, to eliminate redundancy, and to prepare either programs or programming specifications. In addition, the ability to activate a decision-table translator from a typewriter connected to a time-shared computer makes it possible to mechanize much of the paper work of systems analysis and programming.

EXERCISES

1. Give your own definition of the following words, then look up their definitions in *Webster's New International Dictionary*, 2d edition, and compare them with the definitions in the *American Heritage Dictionary*.

    garble, comprise, verbal, infer, cobble, coil, Congressman, up, file, compile, assemble, processor

2. What is the meaning of the following sentence?

    By permitting publication of the Ems dispatch, Bismarck presented to the world a garbled version of the Kaiser's dealings with the French ambassador.

3. What is the meaning of *up* as it is used in the following?

    What's up, Doc?
    He did not live up to the expectations of his family.
    Set up the chessboard while I get some beer.
    Pulling up weeds is hot, dirty work.
    I drew up short when I realized the implications of what I was doing.
    I don't like to get up early.
    If you don't believe me, look it up.
    Situation normal: all fouled up.
    Mary Noble, make up your mind to make up your face, make up your bed, make up a story, and make up with your husband before he makes up to someone else and makes off with her. Make up for lost time or you will make up part of a triangle.

4. How many Congressmen number you among their constituents?

5. What is the meaning of the following sentence?

    The most striking fact about a stored program is that, in operation, it consists of data processing data.

6. The following is a verbal description of an invoicing procedure. Prepare a flowchart and a decision table describing the procedure.

    a. Retail customers do not get a discount and are not shipped goods on consignment. They are required to pay cash on delivery.
    b. No shipment on consignment is made to government agencies. Terms are net 30 days after receipt. No discount on orders totaling

less than $100; 15 percent discount on orders totaling $100 or more.

c. An authorized sales agent for our engines may be shipped engines on consignment. Terms are net 30 days after sale. Discounts are 33 percent for engine orders totaling $50 or more but less than $100; 40 percent for engine orders totaling $100 or more.

d. No consignment shipments of products other than engines are to be made to engine agents. Terms for such products are net 30 days after receipt. Discount is 10 percent.

e. Pumps may be shipped on consignment to authorized pump agents. Discount is 25 percent. Terms are net 30 days after sale.

f. Products other than pumps may not be shipped on consignment to pump agents. Discount on such products is 10 percent. Terms are net 30 days after receipt.

g. No consignment shipments are made to distributors of our products. Terms for pump shipments to an authorized pump distributor are: three equal payments of one third of the total amount to be paid 30, 60, and 90 days after receipt. Discounts are 30 percent for pump orders totaling less than $10, 33 percent for pump orders totaling more than $10 but less than $50.

h. Payment for products other than pumps shipped to a pump distributor is due 30 days after receipt. Discount is 15 percent.

i. Payment for shipments to an authorized fan distributor is due 30 days after receipt. Discounts are: fans, 25 percent; other products, 10 percent.

7. The following is a description of the overtime calculation portion of a payroll procedure. It is a slightly modified version of a real procedure described in a union contract. Prepare a flowchart and a decision table to describe how overtime is to be calculated. Assume that time is reported on daily time cards.

   a. *Definition of terms*

   1) The payroll week shall consist of any seven (7) consecutive days used by the Company for computing the pay of employees (which may or may not coincide with a week beginning at 12:01 A.M. Sunday or at the shift-changing hour nearest to that time).

   2) The workday for the purposes of this Article is the twenty-four (24) hour period beginning with the time the employee begins work, except that a tardy employee's workday shall begin at the time it would have begun had he not been tardy.

   3) The regular rate of pay, as the term is used in section b below, shall mean the hourly rate which the em-

ployee would have received for the work had it been performed during nonovertime hours; for employees on an incentive basis, such regular rate of pay shall be the average straight-time hourly earnings as computed in accordance with existing practices.

b. *Conditions under which overtime rates shall be paid*

1) One and one-half (1½) times the regular rate of pay shall be paid for—
   a) Hours worked in excess of eight (8) hours in a workday.
   b) Hours worked in excess of forty (40) hours in a payroll week.
   c) Hours worked on the sixth (6th) or seventh (7th) workday in a payroll week during which work was performed on five (5) other workdays.
   d) Hours worked on the sixth (6th) or seventh (7th) workday of a seven (7) consecutive day period during which the first five days were worked, whether or not all of such days fall within the same payroll week, except when worked pursuant to schedules described in the sections of this procedure on hours of work; provided, however, that on shift changes the seven-consecutive-day period of 168 consecutive hours may become 152 consecutive hours depending on the change in the shift.
2) Two (2) times the regular rate of pay shall be paid for all hours worked on the holidays specified in the Holidays section of this procedure.
3) Payment of overtime rates shall not be duplicated for the same hours worked. To the extent that hours are compensated for at overtime rates under one provision, they shall not be counted as hours worked in determining overtime under the same or any other provision; provided, however, that a holiday (as established in the section on Holidays), whether worked or not, shall be counted as a day worked in determining overtime under the provisions of section b-1-c above, and hours worked on a holiday shall be counted for purposes of computing overtime under the provisions of section b-1-a above.

FILING YOUR RETURN

You must file a Federal Income Tax Return if:

You are a *U.S. citizen or resident* and your gross income (see Chapter 7 for a definition of gross income) is at least the amount shown below under your particular category (you are considered to be 65 on the day before your 65th birthday):

Single—$1,700 ($2,300 if you are 65 or older);
Married—$2,300 combined income ($2,900 if one spouse is 65 or older, $3,500 if both are 65 or older) and are eligible to file a joint return and are living together at the close of the tax year. See Chapter 2.

However, the filing requirement for each spouse is $600 if:

1) they file separate returns; or
2) they do not have the same household at the end of the year; or
3) another taxpayer is entitled to an exemption for either spouse

You had *uncollected Social Security tax on tips*. See Chapter 9, *Income From Tips*.
You are *self-employed* and had net earnings from self-employment of $400 or more.

The $400 net earnings figure applies to any self-employed individual regardless of age—even to persons 65 or over who receive Social Security benefits.

A U.S. citizen employed in the United States by an international organization, a foreign government, or a wholly-owned instrumentality of a foreign government, and who is exempt from Social Security employee tax, is subject to self-employment tax for his earnings from services performed in the United States. *[Continued on next page]*

FIGURE 3.1 Narrative description: "Who Must File" procedure

You are the executor, administrator, or legal representative of a person who died during the year and, before his death, had satisfied the income requirements for a single or married person, as shown above, or had $400 or more self-employment income;

You are a U.S. citizen living abroad and have met the requirements for single or married persons, as shown above. You may, however, be able to exclude from your gross income a limited amount of earned income:

See Publication 54, available at most U.S. embassies and consulates, or from the Office of International Operations, Internal Revenue Service, Washington, D.C. 20225.

You were a *bona fide resident of Puerto Rico* during the entire tax year and have met the requirements for single or married persons, as shown above. Gross income for this purpose does not include income from sources within Puerto Rico, except for services performed as an employee of the United States or any U.S. agency.

You should file a Federal Income Tax Return if:

You had income tax withheld from your pay but did not have enough income to be required to file a return. By filing a return and claiming your personal exemption, you can get a refund even though you may be claimed as a dependent by another taxpayer.

You are *a survivor* of a decedent who would have been required to file had he lived and no executor, administrator, or legal representative has been appointed.

Illustration. A chart entitled *Who must file* appears on page 137.

FIGURE 3.1 *Continued*

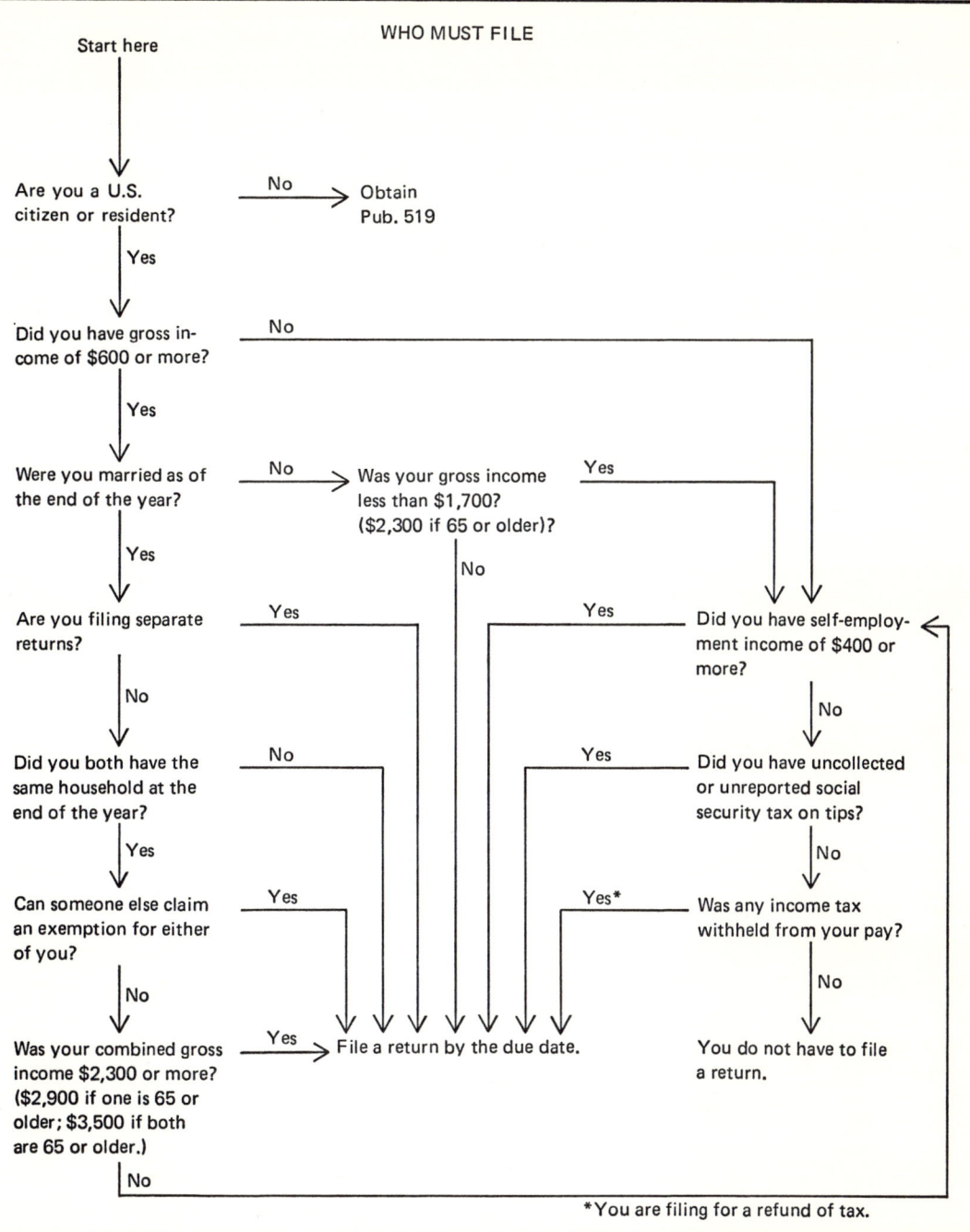

FIGURE 3.2 "Who Must File": Flowchart from Internal Revenue Service booklet.

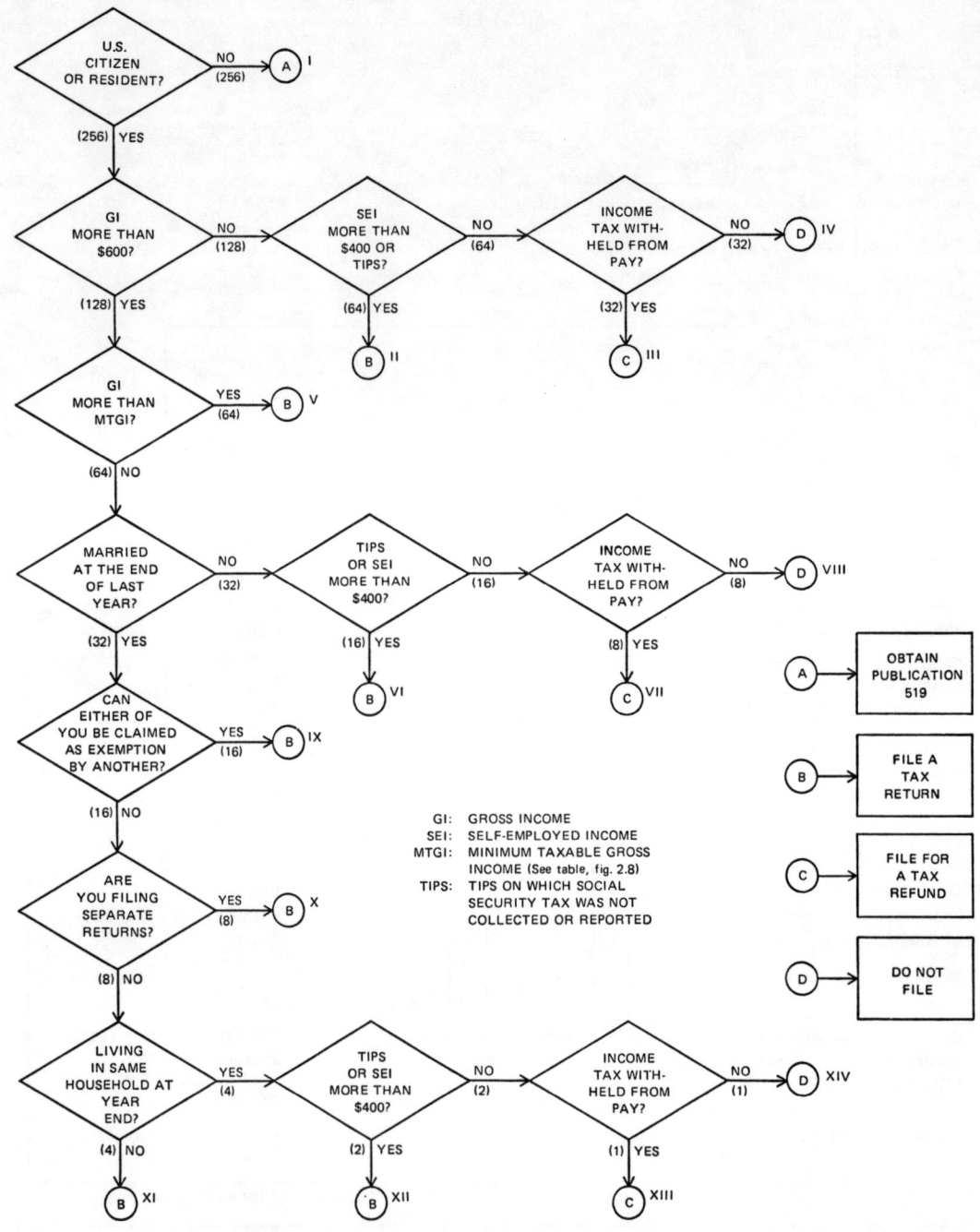

FIGURE 3.3 "Who Must File": Flowchart in standard form (procedure details revised for purposes of exposition)

4

Simple Rules and Composite Rules

A. OBJECTIVES, TERMINOLOGY, DEFINITIONS

Objectives

In chapters 4, 5, and 6, our objectives are—

1. To study the structure of decision tables.
2. To learn how to exploit the characteristics of this structure that make decision tables an effective tool for checking the completeness and consistency of a procedure, eliminating unnecessary portions of a procedure, and transforming a procedure description into a set of programming specifications.

Terminology

To achieve these objectives, we shall have to use language more carefully and precisely than has been necessary in the first two chapters. Up to this point, for example, we have used words like *condition* and *action* and relied on the context to indicate whether we were talking about decision-table *stubs* (in which descriptions of conditions and actions are given) or decision-table *entries* (in which specific *values* or *states* or *specifications* of conditions and actions are given). We have talked about *condition sets* and *action sets* but have not defined them precisely. We have talked about a *correspondence* or *matching up* of the members of a set of sets with the members of another set of sets without explaining in detail what it is that is being matched or placed in correspondence or even what we mean by *matching up* and *correspondence*.

Since our discussion up to this point has been with the externals of a procedure description—that is, its interpretation or meaning—this looseness of expression was employed to avoid considering details that might have distracted attention from the broad outlines of the topics we were discussing. But, at this point, we are no longer concerned with interpretation or meaning; we are concerned with the internal details of decision-table structure. If this concern is not expressed in tight, carefully defined language, we run the risk of confusing rather than enlightening.

Three Fundamental Definitions

Before we define precisely some of the terms we have so far used loosely, let us introduce and discuss three fundamental definitions which will provide a groundwork for all the others.

1. *Set*
   A collection of objects. This definition is clearly circular. It assumes that the meanings of the words *collection* and *objects* are known. This kind of circularity is inescapable in a verbal definition. Only a partial escape from it is possible—an appeal to examples of the defined term. In figure 5.5(a), some examples of sets are given. The reader who encounters the notion of "set" for the first time should examine figure 5.5(a).
2. *Ordered set or vector*
   An ordered set (*vector*) is a collection of objects in which each object conveys information not only by its form or value but also by its position relative to the other objects in the set.
3. *Correspondence*
   A correspondence between two sets is a rule or mechanism which specifies for any member of one set a corresponding member or members in the other set.

Abstract, circular definitions like these do not stick readily in the mind. Again we need examples. One of the most useful examples of an *ordered set* is one which is so familiar to us that we are apt not to recognize its importance and the revolutionary changes it has produced in fields of endeavor as widely separated as pure mathematics and the transactions of everyday life. This example is the ordinary decimal number.

In discussing decimal numbers as examples of ordered sets, let us avoid unnecessary distractions from the central idea by restricting our attention to zero and the positive integers, disregarding fractions and negative numbers. Since there are an infinite number of nonnegative integers, we need not worry about running out of topics of conversation.

It is, indeed, the very fact that a small number of "building-block" symbols makes it possible for us to represent any member of an infinite set that provides one of the most striking illustrations of the usefulness and power of the concept of *vector* or *ordered set.* Decimal numbers use only ten different symbols—the digits 0, 1, 2, 3, 4, 5, 6, 7, 8, 9—to represent any nonnegative integer, *no matter how large it is*. How is this reduction of infinity to a manageable size achieved? It is achieved by associating with each digit not only an *intrinsic* value—which we ascertain by recognizing its shape—but also a value which depends upon position. The *digit* 3, for example, represents the *value* 3 in the number 453 (400 + 50 + 3); it represents the *value* 30 in the number 435 (400 + 30 + 5); it represents the *value* 300 in the number 345 (300 + 40 + 5). And similarly for the other digits in the examples; the values they represent depend both on their shape and on their position in the decimal representation. The "new math"—which has recently caused so many children to question the omniscience of their parents—considers it wasteful to use as many as ten different shapes; it expresses the same set of numbers as ordered sets of only two shapes: 0 and 1.

As an example of what we mean by a *correspondence* between two sets, let us consider the sets *married men* and *married women.* If only monogamous marriages are legal, then to each member of the set of married men there corresponds exactly one member of the set of married women, and conversely. If polygamy and polyandry are legal, then to each member of the set of married men there corresponds one or more members of the set of married women, and conversely. If polygamy is legal, but polyandry is not, then to each member of the set of married men there corresponds one or more members of the set of married women; to each member of the set of married women there corresponds exactly one member of the set of married men.

Second-level Definitions

Using the ideas we have just introduced, let us define some of the technical terms we encountered in the first two chapters:

> *Condition*   A finite set of symbols called *condition values.*
> *Action*   A finite set of symbols called *action values.*
> *Condition stub*   A vector of conditions.
> *Action stub*   A vector of actions.
> *Condition set*   A vector of condition values.
> *Action set*   A vector of action values.
> *Rule*   A correspondence between a condition set and an action set.
> *Entry portion of a decision table*   A vector of rules.

At this point, a reader who has faithfully studied the first two chapters may be pardoned a moment of bewilderment. He has encountered all the defined terms in his prior reading and acquired an understanding of what they mean, yet the definitions given above may well seem incomprehensible to him.

This is because we are changing the ground rules of definition. The definitions of the first two chapters have been definitions by *connotation*, definitions either in terms of the functions performed by the defined objects or in terms of what they mean or suggest to the person reading a procedure description. In this chapter we are shifting to definition by *denotation*, definition in terms of a set of arbitrary undefined elements, independent of external interpretation, and defined within a procedure only by abstract rules that prescribe how they are organized and manipulated.

This transition from external interpretation to internal treatment is fundamental to all of data processing. All a computer can do is process arbitrary collections of digits in arbitrary ways. One of the most widely quoted expressions in computing is "Garbage in, garbage out." This is a picturesque way of saying that incorrect data will produce incorrect answers. But one program's garbage may be another program's bread and butter. If a bank has both savings accounts and a credit-card plan, a sleepy operator may mismatch the inputs of the two programs with the result that what the credit-card holders have paid in interest is reported to the taxing authorities as interest they *earned* on savings accounts.

Our reason for switching from one mode of definition to the other is a reflection of our change from what a procedure *means* to what it *does*. Let us examine the definitions we have just given with this change of outlook in mind.

Condition, action What does it mean to describe a *condition* as a set of *condition values* and an *action* as a set of *action values?* In our very first example of a decision table we encountered a condition called "Make." The decision table in which it appeared was constructed with the understanding that only three makes were to be considered by the procedure: Cord, Reo, Duesenberg. From the standpoint of computer processing this is all we need to know about the "Make" condition—that is, "Make" is defined as three values which are distinguished by three different symbols. When we code the values, "Make" becomes the set of values 1, 2, 3.

Condition stub, action stub Referring to figure 2.3, we see that the condition stub has two conditions: (1) Make (with three values); (2) Condition (with three values). Since both conditions are defined in the code book as the same set of three values (1, 2, 3), how are they to be distinguished from one another?

Clearly, they are to be distinguished by their relative position in the condition stub. *Make* is the first condition; *Condition* is the second condition. In other words, the only identification any particular condition has for our present purposes is its relative position in the condition stub. That is why the condition stub is defined as a *vector* of conditions; the relative position of a condition tells us what it is.

Condition set, action set The short definitions of *condition set* and *action set* given above are correct but incomplete. A complete definition must describe the relation of the condition set to the condition stub, and the action set to the action stub. The complete definitions are as follows:

1. A *condition set* is a vector of condition values. The value in the first position must be selected from the first condition in the condition stub, the value in the second position from the second condition in the condition stub, and so on.
2. Similarly for the values in the *action set;* each must be selected from the corresponding actions in the action stub.

The definitions of *rule* and *entry portion* of a decision table should now present no difficulty. The figures at the end of this chapter introduce us to some of the consequences of our new definitions. They also reflect a viewpoint we stated at the end of chapter 3: our concentration in this book is upon an analysis of the condition-entry portion of a table rather than the stubs or the action entry.

Figure 4.1 illustrates the genesis of our new approach to decision tables. The condition sets are represented in uncoded form. The action sets, however, have been replaced by *action set codes*. Each of these codes represents an *action set* or *course of action*. For the purposes of our analysis, all we need to know about action sets is whether or not they are the same; we are not interested in the specific action values they comprise since we do not intend to use these in our analysis. This does not mean that a complete study of decision tables should not include such an analysis. It should, and some existing decision-table processing programs do incorporate it. But action-set analysis is beyond the scope of this book; our primary concern herein is with condition-set analysis.

Figure 4.1 tells us that several different condition sets lead to the same action set. The condition sets of rules 1 and 4 lead to the action set we have coded A; rules 3 and 6 lead to C; rules 7, 8, 9 lead to E.

B. INTERPRETING COMPOSITE RULES

In figure 4.2 the rules of figure 4.1 have been rearranged so that rules with the same action sets are first placed side by side (figure 4.2a) and then consolidated (figure 4.2b).

The unconsolidated rules are called *simple rules*. They are characterized by the fact that only one condition value appears in each position of the vector representing the condition set. The consolidated rules are called *composite rules*. They are characterized by the fact that, in at least one position, more than one condition value appears. In composite rule I, for example, *Cord* and *Reo* appear in the first position; the same is true of composite rule III. In composite rule V, on the other hand, *all* second-position values (*running well, running poorly, not running*) appear. This is precisely what we mean by a Don't-care entry.

Therefore, instead of listing all the condition values in rule V of figure 4.2b, we make the second entry "Don't care," which means the same thing but takes less writing, particularly when we actually construct decision tables and use the customary convention of expressing "Don't care" by leaving the condition-entry square blank.

The important message conveyed by figure 4.2 is that a composite rule expresses exactly the same relationships as the set of simple rules it replaces. Composite rule I means exactly the same as simple rules 1 and 4; III the same as 3 and 6; V the same as 7, 8, and 9.

Why have composite rules? Because they increase the comprehensibility of a decision table at the same time they reduce its size. We encountered composite rules in the preceding chapters without realizing what they were. In figure 2.8, all the rules with Don't-care entries are composite rules, each standing for a set of simple rules. Rule 1, for example, represents a set of 256 simple rules. How comprehensible would the table be if we had spelled them all out? How big would it be?

C. CONSOLIDATING RULES

Figure 4.2 illustrates how simple rules may be consolidated into composite rules. Figure 4.3 goes further and illustrates how composite rules may themselves be consolidated into more comprehensive composite rules. In both cases, two simple requirements must be met:

1. The rules to be consolidated must specify the same action set.
2. The vectors describing the rules to be consolidated must differ in only one position.

In figure 4.2, rules 1 and 4 differ in the first position. So do rules 3 and 6. Rules 7, 8, and 9 differ in the second position.

In figure 4.3, rules 1, 2, and 3 differ only in the fifth position. So do rules 4, 5, and 6. Both sets of three rules may therefore be consolidated into the two composite rules shown in figure 4.3b. But these two composite rules themselves differ only in the second position. They may therefore be consolidated to give the composite rule displayed in figure 4.3c. This rule is now a substitute for simple rules 1, 2, 3, 4, 5, 6 of the original table displayed in figure 4.3a.

The requirements that must be met before rules can be consolidated are simple but, as is true of many simple things, the details of actually picturing how they work can be confusing to one who does not take the trouble to fix them so firmly in mind that it becomes virtually second nature to apply them. The reader is urged to make sure that he understands clearly how simple rules and composite rules are related. Further, if in the subsequent reading any part of the discussion is hard to follow, there is a good chance that it is because this relation is not as clearly understood as it should be. Should this prove to be the case, the reader is urged to reread this chapter, since most of what follows is based upon the relations we have been discussing.

Note that in figure 4.3 our definition of the decision table is completely abstract. We show five conditions but do not say what they are; they are merely identified as C_1, C_2, C_3, C_4, C_5, with the subscripts (1-5) indicating relative position within the condition stub. For each condition we show the coded set of values that define it. As we noted above, with actions we go even further; we do not specify individual actions but, instead, the codes that identify distinct action sets.

D. EXPANDING COMPOSITE RULES INTO SIMPLE RULES

In figure 4.4, we reverse the process described in the preceding section; we go from a composite rule to the set of simple rules of which it is a consolidation.

In the composite rule used in the example, there are three condition squares that have multiple entries: squares 1, 3, and 5. Square 1 is a Don't-care; this means that all the values specified in the first row of the condition stub are admissible (in this case the values 1 and 2). Square 3 specifies two values explicitly: 2 and 3. Square 5 is again a Don't-care square, meaning that all the C_5 values are permissible; these are 1, 2, and 3.

In figure 4.4b, we see the first stage of the expansion. Condition squares 1-4 (and the action square specifying action set B) are copied exactly as they were in the original table. In the fifth square, however, we make a single entry, one of the values defining C_5. Since there are three of these, we get three rules.

In figure 4.4c, the second stage in the expansion is shown: the expansion of the two values in square 3. This is done by copying the three rules of figure 4.4b exactly as they stand except that single entries are made in condition square 3. Since there are two such entries, we get six rules: three with a 2 in the third position, and three with a 3 in the third position.

The same process is repeated in figure 4.4d. The six rules of 4.4c are copied twice, first with a 1 in the first position, then with a 2 in the first position.

The last expansion leaves no composite rules, so no further expansion is possible.

E. SUMMARY: THE PLACE OF COMPOSITE RULES IN DECISION TABLES

The processes of consolidation and expansion that we have illustrated in the foregoing discussion are fundamental to analyzing "program logic" when it is embodied in decision-table form. As we have seen, composite rules are essential if decision tables are to be both comprehensible and practical. They occur in *all* procedure descriptions: implicitly in narratives and flowcharts; explicitly in decision tables. But they have their price. They introduce the possibilities of inconsistency and redundancy. If inconsistency occurs, the procedure is incorrect; if redundancy occurs, it is inefficient.

Both inconsistency and redundancy can occur when two composite rules overlap. Figure 4.5 gives an example. It shows us the condition portion of two composite rules (referring us to the condition stub of the preceding figures for their exact definition). When we expand the two rules (labeled X and Y) into their constituent sets of simple rules (labeled X_1, X_2 for the first, and Y_1, Y_2, Y_3 for the second), we find that one of the rules making up X (the rule labeled X_1) is the same as one of the rules making up Y (the rule labeled Y_2).

If, in the decision table in which X and Y appear, they both specify the same action set, then no real harm is done; we merely have a relatively harmless redundancy. If, on the other hand, X and Y specify different action sets, we have an inconsistency; we do not know what to do when the vector of condition values (2, 2, 4, 1, 2) occurs in our procedure, since we do not know whether to follow the course of action prescribed for X or that prescribed for Y.

Composite rules also complicate the analysis of the remaining facet of logic in which we are interested: completeness. The next two chapters discuss all three questions (completeness, redundancy, consistency) in greater detail.

A note on usage: What is called a *composite* rule in this text is called a *complex* rule in some other texts on decision tables. Since *composite* suggests something about the nature of the kind of rule it names, it will be the term used throughout this book.

The use of the term *vector* for an ordered set of objects is a little more general than conventional mathematical usage, though many instances of a similar use do occur in mathematical literature. The most common use, however, is to restrict the term *vector* to mean an ordered set of *scalars*— that is, an ordered set in which the only admissible member in any position is a number. An ordered set of vectors is then called a *matrix;* ordered sets of higher dimensionality are called *tensors.*

For our purposes, it is useful to consider that vectors may be made up of ordered sets of quite general objects: *conditions, actions, condition values, action values, rules,* and the like.

The entry portion of a table can thus be called a *vector* of *rules,* which are themselves vectors describing a correspondence between vectors. The essential ideas of vector, as we shall use the term, are—

1. The idea of order or relative position.
2. The idea of set membership.
3. The idea of restricting membership to a specific set of objects.

REVIEW

A good understanding of the *structure* of decision tables is important to anyone who wishes to use them effectively for systems analysis and programming. In order to study this structure, we have to define more precisely some of the technical terms we introduced in the first two chapters. These terms (*condition, action, condition stub, action stub, condition set, action set, rule, entry*) are all defined with the use of three fundamental terms introduced for the first time in this chapter: *set (collection); ordered set (vector); correspondence (matching up)*.

Our study of structure focuses primarily on condition sets and condition entries. Since we are not concerned with the details of action sets (courses of action), we shall ordinarily represent the action portion of a rule by a single square containing the code for a specific action set. Figure 4.1 gives an example of action set coding. The codes enable us to determine easily the facts that are important to our analysis: which rules lead to the same action sets and which to different action sets.

Rules that lead to the same action set can sometimes be combined into one rule. It is at this point that difficulties in analyzing decision-table structure arise. If a decision table consists of a complete set of *simple* rules (every possible condition set placed in correspondence with an action set), its structure is straightforward and easy to understand. But, as we have seen, most realistic decision tables are not of this simple type. Practical decision tables contain many *composite* rules, each such rule replacing a set of simple rules. These composite rules are essential to the practical use of decision tables, yet they complicate the analysis of decision-table structure by making it difficult to determine when a decision table is complete, consistent, and free of redundancy.

Composite rules arise when simple rules are combined. They are characterized by having at least one square in which more than one entry appears. (Remember that a blank square is the same as one in which *all* possible entries appear.) To understand the properties of composite rules, we study two processes: (1) combining rules into composite rules; (2) analyzing (or expanding) composite rules into the sets of simple rules of which they are composed.

Two rules (simple or composite) can be combined into a single rule if they have the same action set and differ only in one condition square. The single rule that replaces the original two will be identical to them except that, in the one square where they differ, the entries from *both* the original rules will appear. By repeated applications of this combining process, simple rules can be combined into composite rules, which can themselves be further combined into higher-level composite rules with two or more multiple-entry squares.

Analyzing, or *expanding*, a composite rule into the set of simple rules that it *covers* or *contains* or *replaces*, is the inverse of the combining process we have just described. It consists of repetitions of one fundamental procedure: locating a multiple-entry condition square and making as many copies of the rule or set of rules in which it occurs as there are entries in the square, each copy identical with the rule or set of rules from which it derives except that single entries appear in what was originally the multiple-entry position.

If we expand the condition portions of two composite rules in the same table, it can happen that the rule sets into which they expand have one or more rules in common. When this is the case, the two rules are said to *overlap*. Complexities in decision-table structure are caused primarily by the occurrence of overlapped rules.

EXERCISES

1. Determine how many distinct action sets there are in the procedure described in exercise 3.6. Identify each with an alphabetic code letter.

2. Do the same for exercise 3.7.

3. Which of the rules in the decision tables of exercises 3.6 and 3.7 are composite rules?

4. Combine the simple rules in the table below into as few composite rules as possible.

C_1:1,2	1	1	1	1	1	1	1	1	1	1	1	1	1	1	1	1	1	1	2	2	2	2	2	2	2	2	2	2	2	2	2	2	2	2	2	2
C_2:1,2,3	1	1	1	1	1	1	2	2	2	2	2	2	3	3	3	3	3	3	1	1	1	1	1	1	2	2	2	2	2	2	3	3	3	3	3	3
C_3:1,2	1	1	1	2	2	2	1	1	1	2	2	2	1	1	1	2	2	2	1	1	1	2	2	2	1	1	1	2	2	2	1	1	1	2	2	2
C_4:1,2,3	1	2	3	1	2	3	1	2	3	1	2	3	1	2	3	1	2	3	1	2	3	1	2	3	1	2	3	1	2	3	1	2	3	1	2	3
Action codes	B	B	B	A	A	C	B	B	B	A	A	D	B	B	B	A	A	E	B	B	B	F	G	C	B	B	B	H	I	J	B	B	B	H	K	L

5. Expand rule 8 of figure 2.8 into its constituent simple rules.

6. In figure 4.4, a composite rule was transformed into a set of simple rules by successive expansions of the multiple entries in rows 5, 3, and 1, in that order. Expand the same rule but change the order of expansion to 1, 3, 5. Do you get the same set of simple rules?

7. Which of the following condition sets overlap?

RULE	1	2	3	4	5	6
C_1: 1,2,3,4,5	2	5	4	3	2	X
C_2: 1,2,3	X	1	2	1	2	1
C_3: 1,2	1	1	X	1	1	X
C_4: 1,2,3,4	3	X	3	X	3	3
C_5: 1,2,3	1	X	1	X	X	1

(X identifies a Don't-care square.)

ENTRY HALF OF FIGURE 2.1 WITH ACTION CODES SUBSTITUTED FOR ACTION SETS

	1	2	3	4	5	6	7	8	9
Condition Sets	Cord	Cord	Cord	Reo	Reo	Reo	Duesen-berg	Duesen-berg	Duesen-berg
	Running well	Running poorly	Not running	Running well	Running poorly	Not running	Running well	Running poorly	Not running
Action Set Codes →	A	B	C	A	D	C	E	E	E

CODES FOR ACTION SETS (Unique Courses of Action)

A: Commission 5%. No shopwork or manager o.k. required.
B: Commission 10%. One week shopwork. No manager o.k. required.
C: Commission 10%. Six weeks shopwork. No manager o.k. required.
D: Commission 10%. Two weeks shopwork. No manager o.k. required.
E: Variable commission. Shopwork to be estimated. Manager o.k. required.

FIGURE 4.1 Condition sets and coded action sets

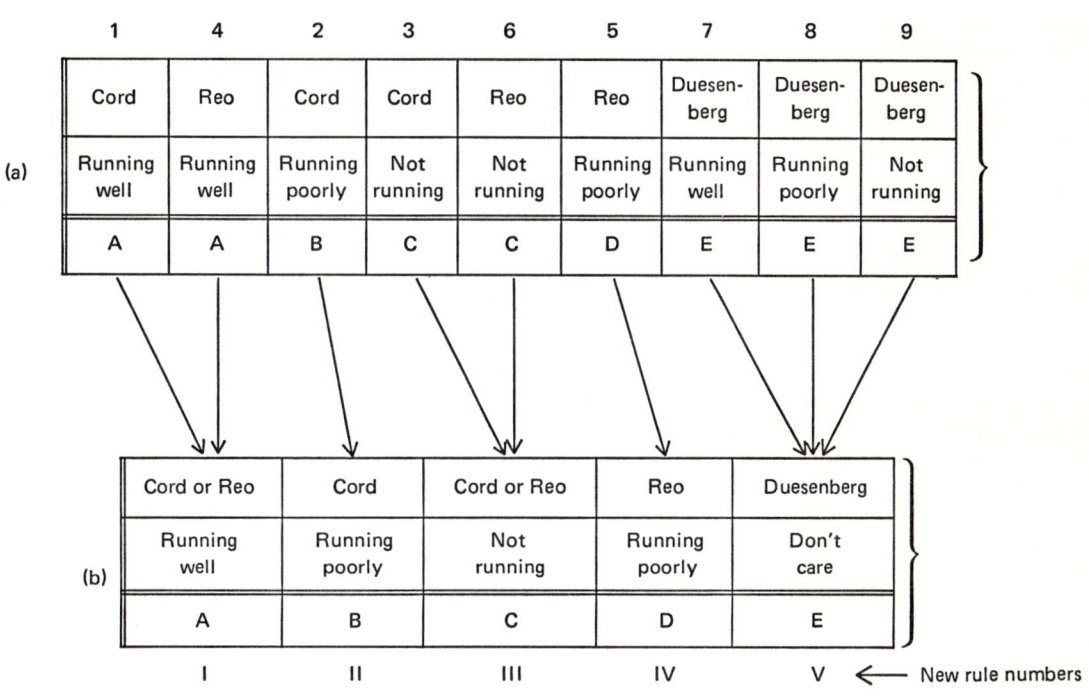

FIGURE 4.2 Consolidating simple rules into composite rules: I

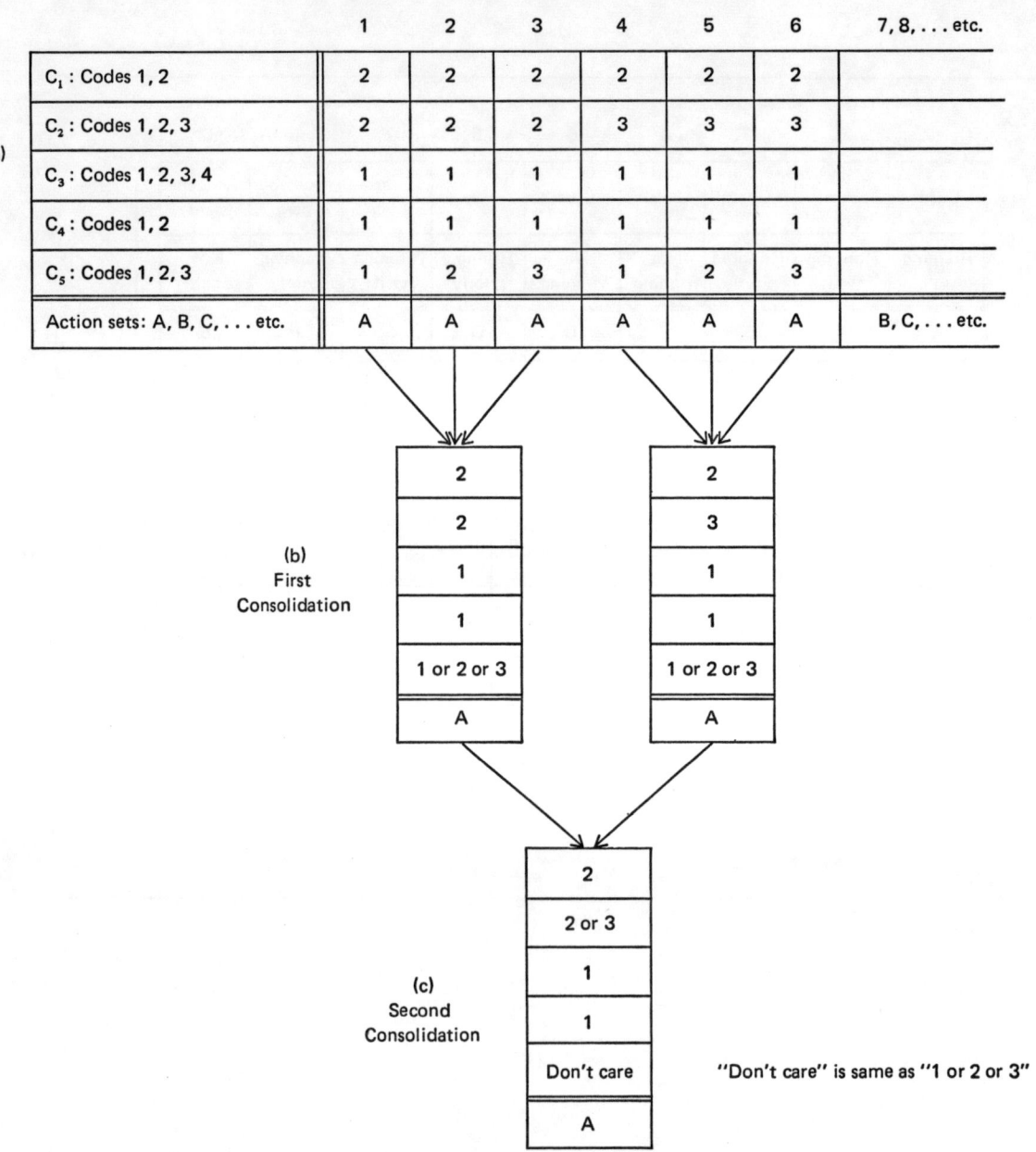

FIGURE 4.3 Consolidating simple rules into composite rules: II

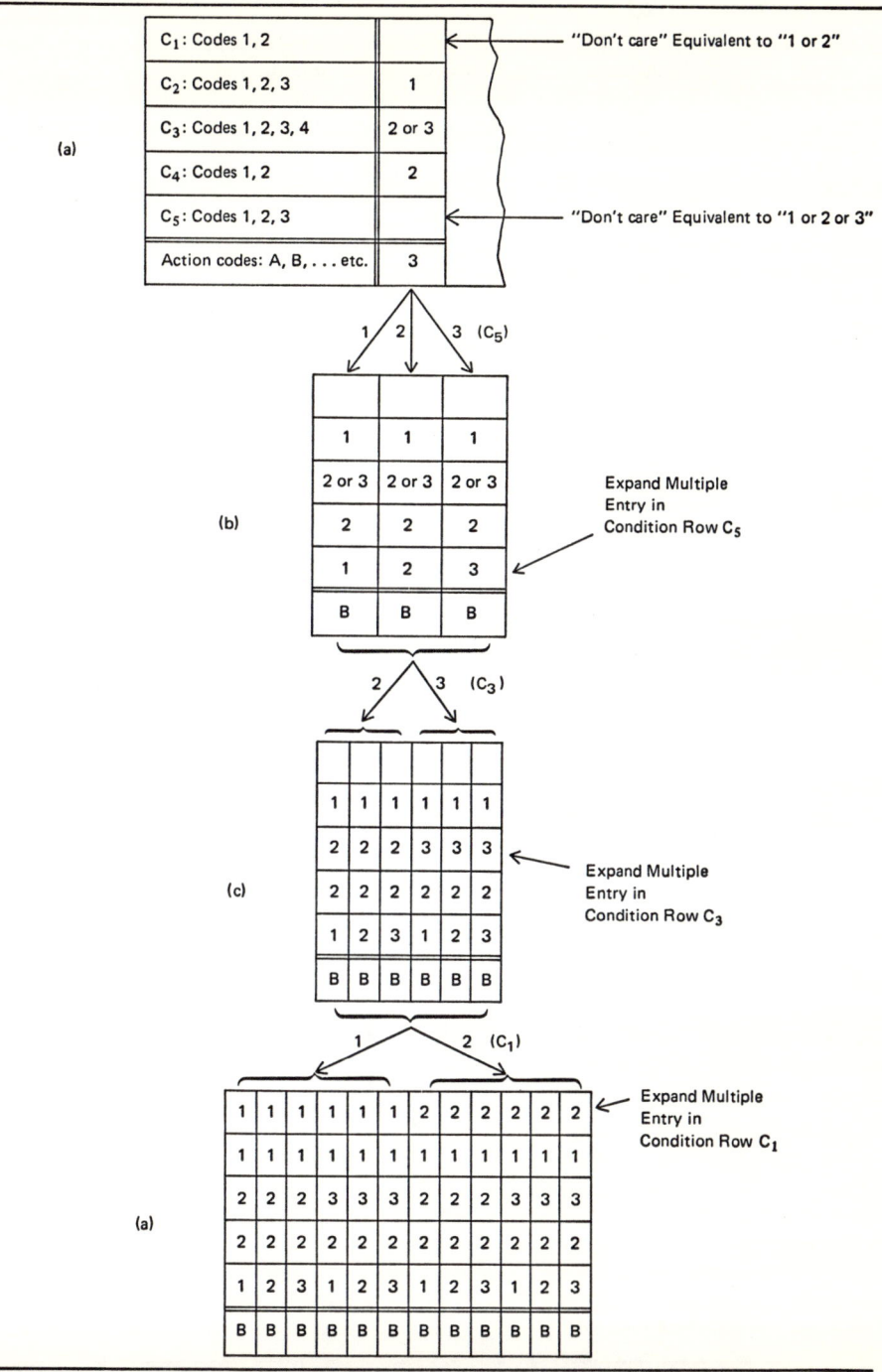

FIGURE 4.4 Expanding a composite rule into its constituent set of simple rules

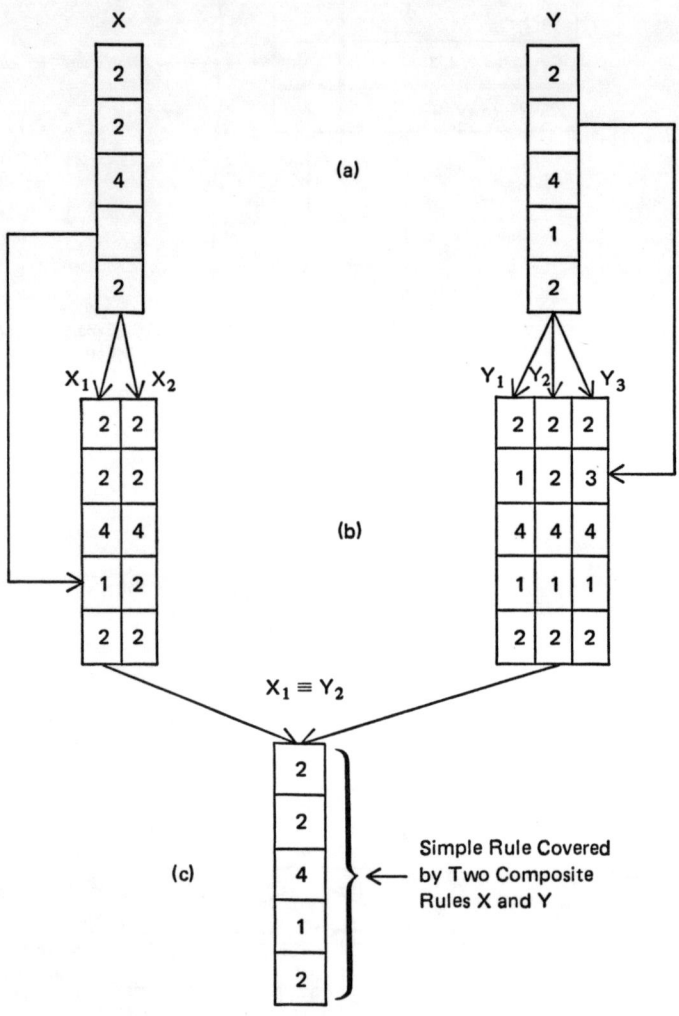

FIGURE 4.5 Composite rules that overlap

5

Completeness and Consistency I

A. HAVE WE CONSIDERED ALL POSSIBILITIES?

We are now in a position to consider the question When is a decision table *complete*?—that is, when can we be sure that we have specified an action set for each possible condition set?

Before we can answer this question, we have to answer two others:

1. *How many* distinct condition sets can we write if we restrict ourselves to the conditions specified in a given condition stub?
2. *What are* those condition sets?

Figure 5.1 displays a condition stub and a calculation that tells us how many distinct condition sets can be derived from that stub. Figure 5.2 shows us what those condition sets are by displaying them as a collection of 144 five-digit numbers in which the leftmost position of a number corresponds to the top condition square, the one to its right corresponds to the square just below the top, and so on, the digit positions from left to right corresponding to the condition-square positions from top to bottom.

We have thus "digitized" the condition-entry portion of a complete table. At this point, a hardheaded, realistic systems analyst might say, "So what? You show me a bunch of numbers on a page and expect to impress me. Why? What's the use of a collection of numbers? I don't want theory. I want something practical. Are those numbers you're discussing practical?"

The answer is that they're as practical as a parts catalog, a price book, a "data base" system, the location or bin numbers in a warehouse, a customer or vendor identification file, or any other of the enormous variety of code books identifying the significant features of a business: for example, the codes for finished and semifinished products, cost and expense, plant, property and equipment, sales analysis, freight rates, personnel statistics, tax and accountability categories, work orders, purchase orders, and so on. The code combinations we're concerned with exist right now. Usually, they've been devised piecemeal over a period of years. Frequently, the way they're constructed reflects the influence of history—perhaps the history of a company's early attempts to mechanize its procedures by using what used to be called *punch-card* or *tab* equipment and is now called *unit record* equipment. This structure of the codes is likely to persist even though it may be outdated because the unit record equipment has been replaced by a stored-program computer system for which an alternative structure would be more efficient and informative.

What we are doing in this and the following chapter is considering *explicitly* what is implicit and frequently unrecognized in the systems and procedure activities of many companies: the properties of the coding structure that describes what an organization is, what it does, and how it does it. Effective control of a modern organization is possible only if information about how it is behaving is readily available to its managers. The validity, timeliness, accuracy, and relevance of the reports that present this information depend directly upon the adequacy of the coding structure that controls the preparation of the reports. If sales-analysis codes are not properly devised, the company's marketing program can be in trouble long before any indication of this trouble surfaces in a periodic report. If product-category codes are too broad, they may not provide useful information about the profitability of specific product lines; if they are too narrow, the profitability calculations based on them may be invalid.

Coding structures are fundamental to systems analysis. A major advantage of the decision table is that it permits us to discuss systems and procedures directly in terms of the coding structures on which they are based.

B. THE NUMBER OF SIMPLE RULES IN A TABLE

The numbers displayed in figure 5.2 look like the numbers of ordinary arithmetic. Can this similarity be exploited?

The answer is yes. The question "How many simple rules are there in a table with five condition rows?" is the same kind of question as "How many numbers can be written with exactly five digits?"

In the latter case, we know that the numbers go from 00000 to 99999 and include every integer in-between. There are 99999 positive integers in the range 1–99999. Adding 1 to account for the number zero, we get $1 + 99999 = 100{,}000$ as the total number of distinct, five-digit decimal-number representations.

Unfortunately, calculating the total possible distinct code combinations in a decision table is not quite as straightforward. The first number in our table is 11111; the last is 23423. There is no obvious way to deduce, from these two numbers, the fact that there are 144 combinations in the set that begins with one and ends with the other.

Let us go back to first principles. How many different decimal numbers can we write with just one digit? Obviously, ten—the digits themselves; 0, 1, 2, 3, 4, 5, 6, 7, 8, 9. How many different numbers can we write with two digits? In the second position from the right we can again write one of ten digits. We can follow each of these with the ten one-digit numbers listed above. (That is why the second position from the right was called the *tens* position when we studied elementary arithmetic.) The total number of combinations of first and second digits is thus $10 \times 10 = 100$. If we go to the third, or *hundreds*, position, we can again write ten digits, but now each of these can be followed by one hundred two-digit combinations; the total number of possibilities is thus $10 \times 100 = 10 \times 10 \times 10 = 1000$. And similarly for the *thousands* position ($10 \times 1000 = 10,000$) and the *ten-thousands* position ($10 \times 10,000 = 100,000$).

The reasoning for the five-digit code numbers of figure 5.2 is exactly the same, but the number of different digits permitted in each position is no longer a fixed value, 10, as it is in the decimal number. Instead, it is the number of distinct values in the corresponding position of the condition stub. In our example, these numbers are 2, 3, 4, 2, 3. Therefore, in the units position, we can now write only one of three digits; in the position to its left, we can write only one of two, with each of which we can associate the three units-position values, for a total of $2 \times 3 = 6$ combinations; in the third position from the right, we can write one of four digits, and with each of these we can associate six two-digit combinations, for a total of $4 \times 6 = 24$ three-digit combinations; and so on through $3 \times 24 = 72$ and $2 \times 72 = 144$. In our code combinations, the value of the positions proceeding from right to left is no longer units, tens, hundreds, thousands, ten-thousands, as it is in the decimal number; instead it is units, threes, sixes, twenty-fours, seventy-twos.

Figure 5.3 illustrates the parallelism between the construction of five-digit decimal numbers and the construction of five-digit codes for our example.

The relationships are easier to see than they are to describe. In figure 5.2, we see that the rightmost position cycles through the numbers 1, 2, 3 throughout the table. Two digits (1, 2) can occupy the position immediately to its left. Each of these is held constant for a units-position cycle. The result is six two-digit codes which, in turn, cycle throughout the table. Similar rules govern the cycling of the three-, four-, and give-digit combinations. The last is, of course, "cycled" only once.

The justification for the calculation in figure 5.1 should now be apparent: Each condition in a condition stub is a set of two or more values. To calculate the total number of rules in a table, write down the numbers that tell us how many values there are in the first condition, the second, the third, and so on for all the conditions in the stub. Multiply these numbers together. The resulting product is the total number of simple rules in the table.

C. THE NUMBER OF SIMPLE RULES IN A COMPOSITE RULE

Calculating the number of simple rules in a composite rule is just like calculating the number of simple rules in a table. At least one condition square in a composite rule has more than one value in it. The others may each have one or more. If we write down the number of values in each square and multiply them all together, we get the total number of simple rules covered by the composite rule.

An example is shown in figure 5.4, in which we calculate the number of rules in the composite rule we expanded in figure 4.4. Happily, our answer, 12, agrees with the number of simple rules we show at the bottom of figure 4.4.

The determination of what the simple rules in a composite rule actually are is similar to the determination of the simple rules in a complete table. The only difference is that some squares have only one value. In the positions corresponding to those squares, these values will, of course, be the only ones cycled—that is, they will be the same for all the simple rules. The others cycle through their values in the manner described for a complete table. Figure 4.4 shows us how the cycling works for this particular example.

D. ELEMENTARY IDEAS FROM THE THEORY OF SETS

This book is intended primarily for the nonmathematical reader. The mathematician who has read the foregoing has already written the formula for the number of rules in a complete table with k conditions as $n_T = n_1 \times n_2 \times \ldots \times n_k$. For the number of simple rules in a composite rule, he has probably written a product of doubly subscripted factors supplemented by index information about the second index in each factor.

Such symbolic formulations of the processes we are describing are useful, desirable, even essential—for the needs of a mathematician. For the nonmathematician, they are apt to confuse, terrify, or bore rather than inform. For this reason, they are avoided throughout most of this book.

For some purposes, however, this avoidance of explicit mathematical ideas is undesirable. This is clearly true when it is less confusing to discuss underlying mathematical ideas than it is to discuss a specific topic without reference to the mathematics that describes it.

We are now ready to discuss the structure of overlapped composite rules—potential sources of either inconsistency or redundancy. This discussion will be a great deal more comprehensible if we first familiarize ourselves with some elementary ideas from the mathematical theory of sets.

Most of the ideas we need are illustrated in figure 5.5.

Figure 5.5a provides some examples of sets. Note that the idea of set is quite general. The sets given as examples include car makes, car running conditions, colors, fruit, numbers, and the set of words (YES, NO). The same entity may be a member of several sets; the entity 1, for example, is a member of all the sets of numbers given as examples. In our usage, sets are represented as lists of objects separated by commas and enclosed in parentheses.

One of the most important ideas of set theory, for our purposes, is that of *set overlap*. The overlap of two sets is the set made up of the members they have in common. In set theory, the words *intersection* or *meet* of two sets are used to denote the subset that two sets have in common. These *overlap* or *intersection* or *meet* subsets are shown to the right of the equals sign in the examples displayed in figure 5.5b.

For some purposes, we wish to consider not just the elements that two sets have in common but all the distinct elements that occur in both sets. This is called the *union* or *join* of the sets. Again, the examples (figure 5.5c) help explain the idea. For the purposes of this book, this concept is not as important as the concept of *overlap*.

Figure 5.5d introduces three important ideas: *complementary* set, *universal* or *full* set, *empty* or *null* set. These merit further discussion.

A *full* set is the set of all values. What these values actually *are* depends upon what set we are considering. Most of the sets important to us are the conditions in the condition stub of a decision table. Each condition is the *universe* or *universal set* or *full set* for its row. This is merely another way of saying that the sets in the condition squares along any row must be made up of members taken from the condition in the same row.

The *empty* or *null set* is the set with no members. This set will be of particular importance when we come to consider whether two composite rules overlap. When two sets have no members in common, their overlap is the *empty* set—the set with no members.

The third idea, *complementary set,* will be important when we wish to describe sets not by listing what members they include but by listing what members they don't include. If a full set consists of the integers from 1 to 15, for example, and we wish to denote a subset which consists of all those integers except 7, it is more convenient to write (∼7) than (1, 2, 3, 4, 5, 6, 8, 9, 10, 11, 12, 13, 14, 15). Note that the complement of the full set is the empty set and the complement of the empty set is the full set. Note further that, in decision tables, a don't-care entry in a condition square specifies the full set for that square.

One last technical term. When two sets have no members in common, they are said to be *disjoint*. We shall use this term both for sets and for composite rules. A composite rule is a set of simple rules. When two composite rules have no simple rule in common—that is, when they do not overlap—we shall say that they are *disjoint*.

E. WHEN COMPOSITE RULES OVERLAP

Let us apply the ideas of the preceding section to the analysis of composite rules. As a reflection of our new approach, let us generalize our definition of condition set slightly. In chapter 4, we said it was a vector of condition values. This is adequate when we are considering only *simple* rules, since a simple rule has exactly one value in each condition square. If we wish to include *composite* rules in our definition, we must generalize it as follows:

> *Generalized Definition of Condition Set:* A condition set is a vector of *sets* of condition values, the set in any position being a subset selected from the condition in the corresponding position in the condition row. A *subset* of a given set is a selection of members from that set. The selection can include from zero elements selected from the set (the empty set) to all the elements of the set (the full set).

In terms of this new definition, a *simple* rule becomes a rule each of whose condition-square subsets contain exactly one member. A *composite* rule becomes a rule each of whose condition-square subsets contains at least one member, and one or more of whose subsets contains more than one member. A complete *condition entry* is a rule all of whose subsets are full sets. An *empty* or *null* rule is a rule in which at least one condition-square subset is empty. (An empty or, as we shall usually call it, a *null* rule is important when we consider rule overlap.)

We apply these new ideas in figures 5.6 and 5.7.

Figure 5.6 shows the expansion of two composite rules based on the condition stub of figure 4.3. The sets of rules described by the two composite rules have four members in common. Using the "digitized" description of condition sets, we can describe these rules as 11221, 11222, 11321, 11322.

The advantage of introducing the ideas of set theory is illustrated in figure 5.7. The four rules that the two composite rules have in common are the *overlap set of rules*. In figure 5.6 we determine what this overlap set is by expanding each composite rule into a set of simple rules and seeing what members the two expansions have in common. The result is a subset of four rules. This subset can be described by a single composite rule. In figure 5.7, we determine this composite rule directly—that is, without doing any expansions—by comparing corresponding condition squares and seeing where they overlap.

Let us go, step by step, through the comparison of rules in figure 5.7, comparing corresponding positions in rule A, taken from figure 5.6a and rule B, taken from figure 5.6b.

In the first position, rule A has the full set, and rule B a set consisting of one member: (1). The overlap is the subset (1).

In the second position, both rules have a subset with one member: (1). Again the overlap is the set with 1 as its only member.

In the third position, A has (2, 3) and B has the full set. The overlap is (2, 3).

In the fourth position, the overlap is (2); in the fifth it is (1, 2).

The overlap rule is thus the composite rule shown in figure 5.7b. In digitized form, we can display it as 11(2, 3)2(1, 2). This is shorthand for 11221, 11222, 11321, 11322. Comparing this with the explicit display in figure 5.6c, we see that the single composite rule we have generated covers the four rules that are shown in figure 5.6c.

These ideas provide the basic foundation for the discussion of completeness and consistency in the next chapter.

F. PRACTICAL ASPECTS OF A THEORY OF CODING

The man who considers himself practical frequently shies away from theory. He usually regards it as something that belongs in a school rather than in a factory or an office. He does this despite the fact that all aspects of his daily life are shaped by abstract theories which, finding expression in a bewildering rate of technological expansion, determine how and what he eats, how he constructs his habitation, how he gets from one place to another, what kind of entertainment and recreational activities he can enjoy, how he uses the earth's resources; in short, how he transforms the world in which he lives—and how it transforms him.

Because of this scorn of theory by "practical" men, the technology of gadgetry is rapidly outstripping the technology of ideas or concepts. Our ability to build things is far greater than our ability to use them properly. This will continue to be true as long as we overestimate the importance of developing gadgets and underestimate the importance of developing ideas—as long as we are more impressed with the speed with which tapes move or disks revolve or console lights blink than we are with the ideas that control the tapes, disks, and lights.

Our ability to understand and control an organization depends directly upon how well we understand its structure. The only way to achieve this understanding is to analyze the structure into a set of fundamental elements and describe how these elements interact. For most modern organizations, the number of these elements is so great and the interactions are so intricate that they cannot be visualized without the use of modern "data-processing" or "information-processing" machines. These machines, however, can be misused. The utility and validity of the reports they generate depend directly on the description of organizational structure that is built into the programs that control them. For digital computers, the most efficient way to describe structure is by means of digital codes—the kind of codes we have been discussing.

REVIEW

In order to determine whether a decision table is complete, we have to know how many different simple rules can be generated from the conditions in its condition stub.

This can be done most easily by determining how many different digital codes we can generate from the coded condition values in the condition stub. The codes will have as many digit positions as there are conditions in the condition stub; in each digit position, only values from the condition in the corresponding position are permitted. The result is that we can calculate the number of different code combinations in the same way we calculate the number of different decimal numbers that can be written with a specified number of digits. In the case of the decimal numbers, any one of ten digits is possible in any position. To determine how many different five-digit decimal numbers there are, for example, we multiply together five tens ($10 \times 10 \times 10 \times 10 \times 10$) and get the number 100,000.

In a decision-table code, instead of ten possible symbols for each position, we have the number of values in the condition for that position. If these numbers are 2, 3, 4, 2, 3, we multiply these numbers together and get ($2 \times 3 \times 4 \times 2 \times 3$) = 144.

To generate all 144 combinations for this example, we start out with the initial values in each position (11111). We then let the rightmost position cycle through the numbers 1, 2, 3. For the next code, we advance the second position from the right to 2 and allow the rightmost to recycle through the values 1, 2, 3. This same process is repeated for all positions from right to left, each position being reset after it and all the positions to its right have reached their maximum value.

Similar methods apply to the determination of the number of simple rules in a composite rule and the generation of the simple rules themselves.

To study redundancy and consistency, we must learn to detect when two composite rules cover one or more of the same simple rules. This can most conveniently be done by using some of the ideas from the theory of sets.

The most important idea, for our purposes, is the idea of set *overlap*. Two sets of objects overlap when one or more objects or entities are members of both sets. The *overlap* set of two or more sets is the set of members they all have in common.

Other important ideas are *complement* or *complementary set* (the set of all members *not* included in a given set); *universe* or *universal* or *full* set (the set of all objects being considered); *null* set (the set with no members); *combined set* (the set of distinct objects that appear in one or more sets being combined).

One way to determine whether two composite rules overlap is to generate, for each one, the set of simple rules it covers. If at least one simple rule occurs in both sets, the two composite rules overlap. Using ideas from the theory of sets, we can avoid this laborious process by determining an overlap rule directly. If this rule has an empty set in one or more of its squares, we know there is no overlap. If not, we not only know that overlap exists, we also have a composite rule that describes the set of overlapped simple rules.

EXERCISES

1. For the purposes of arithmetic, numbers are usually represented in terms of powers of a fixed *radix* or base: 10, 2, 8, 16, 7, etc. The powers of the base 10 ($10^0 = 1$, $10^1 = 10$, $10^2 = 100$, . . .) give the representation with which we are most familiar.

    The numbers that represent coded limited-entry condition sets are *fixed-base* or *fixed-radix* representations in the base 2—that is, binary numbers.

    Numbers in a coded extended-entry table are, in general, not represented in powers of a fixed base. Instead, they are *mixed-radix* representations.

    This sounds unusual and unfamiliar, yet mixed-radix representations are part of our daily life. They are usually used in the measurement of time, length, area, weight, and the like. The following examples illustrate a few of these common mixed-radix uses:

    a. A 24-hour digital clock gives the time in hours, minutes, and seconds. How many different times can it display?
    b. How many combinations of feet and inches are less than a yard?
    c. The smallest unit in the apothecary system of weight measurement is a grain. There are—

    > 20 grains in a scruple
    > 3 scruples in a dram
    > 8 drams in an ounce
    > 12 ounces in a pound

    How many combinations of ounces, drams, scruples, and grains are less than an apothecary pound? How many grains are there in a pound?

2. In a crossword puzzle, we have determined that the second, fourth, and fifth letters are, respectively, R, N, T. Assuming that the third letter is a vowel, how many different five-letter combinations does the would-be puzzle solver have to try if he wants to be sure he has covered all possibilities?

3. The numbered rules in the table below describe a list of words. The "Else" rule describes letter combinations that, for the purpose of this table, are not considered words.
   a. What are the words described by the table?
   b. How many combinations are not considered words?

C1: C, D, F, G	C	C	C	C	C	D	D	D	D	F	F	F	F	G	G	G	G	ELSE
C2: A, O, U	A	O	O	U	U	A	A	O	U	A	O	O	U	U	A	A	O	U
C3: L, M, P, R	M,P,R	M,P,R	P	R	P	M	L,R	L,M,P	P	M,R	P	R	M	R	L	M,P	R	R M
C4: E, S	E,S	E	S	E,S	S	E,S	E	E	E	S	E	E	S	E	E,S	S	E	S

4. The *universe* or *universal set* or *full set* for this exercise is the set (BAD, BED, BUD, BAR, BAN, BUN, CAB, CAD, CAN, CAR, CUR, FAD, FAN, FAR, FEW, FUN).
   a. What is the subset of words that contain an A in their spelling?
   b. What is the subset of words that contain an F in their spelling?
   c. What is the subset that contains an E?
   d. What is the subset that contains either an E *or* an F?
   e. What is the subset that contains both an E *and* an F?
   f. What is the overlap set of the subsets defined by *a* and *d* above?
   g. What is the overlap set of the subsets defined by *a* and *c* above?

5. The two sets for this exercise are the sets of letters

   (M,E,A,N) , (M,A,N,Y)

   a. What is the overlap of these sets?
   b. What is their combination (union, join)?

6. Is the decision table of figure 2.8 complete? Do any of the rules overlap?

7. The decision table of figure 2.2 is clearly incomplete.
   a. How many rules are missing?
   b. How should the table be completed?

Condition Stub of Figure 4.3	Total Number of Valid Values for Each Condition
C_1: Codes 1, 2	2
C_2: Codes 1, 2, 3	3
C_3: Codes 1, 2, 3, 4	4
C_4: Codes 1, 2	2
C_5: Codes 1, 2, 3	3

Total Number of Rules in Complete Table
= 2 x 3 x 4 x 2 x 3 = 144

FIGURE 5.1 Calculating the total number of rules in a complete decision table

11111	11112	11113	11121	11122	11123
11211	11212	11213	11221	11222	11223
11311	11312	11313	11321	11322	11323
11411	11412	11413	11421	11422	11423
12111	12112	12113	12121	12122	12123
12211	12212	12213	12221	12222	12223
12311	12312	12313	12321	12322	12323
12411	12412	12413	12421	12422	12423
13111	13112	13113	13121	13122	13123
13211	13212	13213	13221	13222	13223
13311	13312	13313	13321	13322	13323
13411	13412	13413	13421	13422	13423
21111	21112	21113	21121	21122	21123
21211	21212	21213	21221	21222	21223
21311	21312	21313	21321	21322	21323
21411	21412	21413	21421	21422	21423
22111	22112	22113	22121	22122	22123
22211	22212	22213	22221	22222	22223
22311	22312	22313	22321	22322	22323
22411	22412	22413	22421	22422	22423
23111	23112	23113	23121	23122	23123
23211	23212	23213	23221	23222	23223
23311	23312	23313	23321	23322	23323
23411	23412	23413	23421	23422	23423

FIGURE 5.2 Complete display of condition combinations (condition sets, condition vectors) for condition stub of Figure 5.1

DECIMAL NUMBERS

Position Name	Number of Permissible Digits	Position Value	Number of Possible N-digit Combinations
Units	10	1	10 x 1 = 10 1-digit combinations
Tens	10	10	10 x 10 = 100 2-digit combinations
Hundreds	10	100	10 x 100 = 1000 3-digit combinations
Thousands	10	1000	10 x 1000 = 10,000 4-digit combinations
Ten-thousands	10	10000	10 x 10,000 = 100,000 5-digit combinations

CODE NUMBERS

Position Name	Number of Permissible Digits	Position Value	Number of Possible N-digit Combinations
Units	3	1	3 x 1 = 3 1-digit combinations
Threes	2	3	2 x 3 = 6 2-digit combinations
Sixes	4	6	4 x 6 = 24 3-digit combinations
Twenty-fours	3	24	3 x 24 = 72 4-digit combinations
Seventy-twos	2	72	2 x 72 = 144 5-digit combinations

Note that, in both cases, the Position Value is determined by the number of possible digit combinations in the positions to the right.

FIGURE 5.3 Parallelism between the construction of five-digit decimal numbers and the construction of five-digit codes

Composite Rule (Figure 4.4)	Number of Permissible Entries in Condition Squares
1 or 2	2
1	1
2 or 3	2
2	1
1 or 2 or 3	3

Total Number of Constituent Simple Rules
= 2 x 1 x 2 x 1 x 3 = 12
(Compare with Figure 4.4)

FIGURE 5.4 Calculating the total number of simple rules making up a composite rule

(a) EXAMPLES OF SETS

(Cord, Reo, Duesenberg), (Running well, Running poorly, Not running),
(Yes, No), (0,1), (1,2,3,4), (1,4), (1),
(Apples, Oranges), (Black, Blue)

(b) SET OVERLAP (Logical terminology: Intersection, Meet)

Symbol: ∩

(1,2,3,4) ∩ (1,2,3) = (1,2,3)
(1,2,3,4) ∩ (1,4) = (1,4)
(1,2) ∩ (2,3,4) = (2)
(Cord, Reo) ∩ (Reo, Duesenberg) = (Reo)
(1,2,3) ∩ (1,2,3) = (1,2,3)

(c) SET COMBINATION (Logical terminology: Union, Join)

Symbol: ∪

(1,2,3,4) ∪ (1,2,3) = (1,2,3,4)
(1,2,3,4) ∪ (1,4) = (1,2,3,4)
(1,2) ∪ (2,3,4) = (1,2,3,4)
(1,2,3) ∪ (1,2,3) = (1,2,3)
(Cord, Reo) ∪ (Reo, Duesenberg) = (Cord, Reo, Duesenberg)

(d) SET COMPLEMENTATION (Logical terminology: Complementation, Negation)

Symbol: ~

Note: To complement a set, we must know the full range of permissible values: the 'universe of discourse'.

Universe   Sample Complementary Relationships

(0,1):   ~ (0)   = (1)
         ~ (1)   = (0)
         ~ (0,1) = ( ), The 'empty' or 'null' set.
                   This is sometimes symbolized: (ϕ)
         ~(ϕ) = (0,1), The 'full' set or 'universe'

(1,2,3,4):   ~ (1)   = (2,3,4)
             ~ (1,4)   = (2,3)
             ~ (2,3)   = (1,4)
             ~ (ϕ) = (1,2,3,4)
             ~ (1,2,3,4) = (ϕ)

FIGURE 5.5 Ideas from set theory: overlapping, combining, complementing

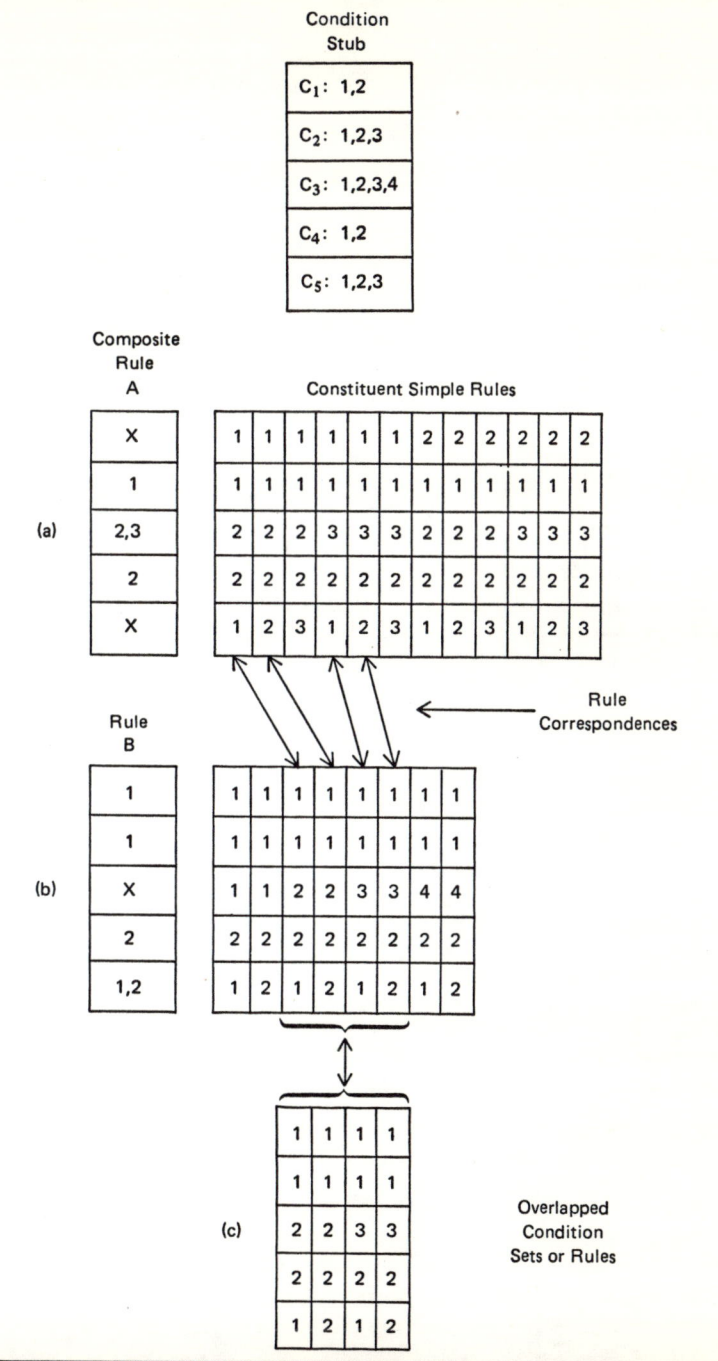

FIGURE 5.6 Example of rules that overlap

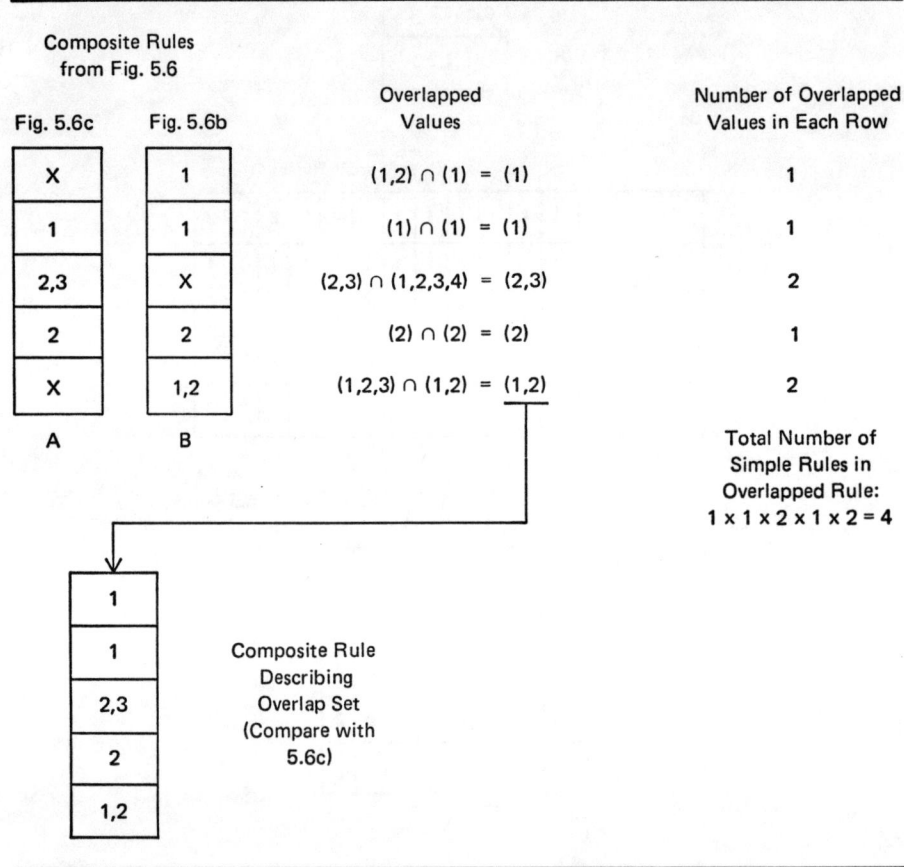

FIGURE 5.7 Calculating the overlapped condition sets in two composite rules

6

Completeness and Consistency II

A. IS THIS RULE NECESSARY? ARE THESE RULES COMPATIBLE? IS THIS RULE POSSIBLE?

In this chapter, we are concerned with three questions:

1. *Redundancy*   Does a given table contain any unnecessary rules?
2. *Interrule consistency*   Does any rule in a table contradict other rules?
3. *Intrarule consistency*   Does any rule in a table include an impossible condition set?

The discussion will be brief. For practical tables, these questions are more easily answered by a program than by a person. It is beyond the scope of this book to describe the details of such a program. We shall merely sketch some of its general characteristics.

B. TYPES OF REDUNDANCY

Two rules are redundant if they have the same action set and if their condition sets overlap.

If two rules are exactly the same, one is obviously redundant. If two rules are different, we calculate their overlap rule. We now have two possibilities:

1. The overlap rule is the same as one of the original rules.
2. The overlap rule is different from both of the original rules.

Figure 6.1 illustrates an example of the first possibility. The overlap rule is exactly the same as rule B. Rule B is therefore unnecessary, since all of its simple rules are contained in rule A.

Figure 4.5 illustrated a different kind of overlap. Rule X contains two simple rules (X_1, X_2). Rule Y contains three simple rules (Y_1, Y_2, Y_3). X_1 and Y_2 are the same, so we have a redundancy, since the same rule appears in two places in the table. This is different from the complete redundancy of rule B in the previous example, however, since we cannot eliminate it by merely dropping rule X or rule Y. If we dropped either one, we would be dropping one or more rules that are *not* redundant.

C. CHECKING FOR COMPLETENESS; REMOVING THE OVERLAP

When two rules overlap, but neither is completely redundant, we have a situation that complicates the procedure for checking a table for completeness. Here is the easiest way to check for completeness:

1. Calculate the number of rules in a complete table (section 5B).
2. Calculate the number of rules in the condition entry. To do this, count each simple rule as 1, count each composite rule as the number of simple rules it contains (section 5C), and add up the counts for all the rules.
3. The table is complete if the sum we calculate in step 2 is the same as the product we calculate in step 1.

This procedure is valid only if no rules overlap. If rules overlap, we are, in effect, counting some simple rules more than once. This would make the completeness checking procedure invalid.

What we need is a procedure for removing rule overlap—that is, a procedure for replacing two overlapping rules by a set of rules which do not overlap but which cover the same set of distinct simple rules.

A simple case of removing rule overlap is illustrated in figure 6.2. Two composite rules, A (covering 12 simple rules) and B (covering 8 simple rules), have an overlap rule that contains 4 simple rules. The total number of *distinct* rules covered by A and B is therefore *not* $12 + 8 = 20$ but, instead, $12 + 8 - 4 = 16$, where, in the second calculation, we subtract the number of simple rules that were counted twice when we added the number of rules in A and B.

Can we modify either A or B so as to eliminate the overlap but still cover the same set of distinct simple rules? Yes, there is a simple way to modify rule B so that it no longer includes the overlap rule but continues to include all of its original rules that are not covered by the overlap rule.

We do this by noting that rule B is exactly the same as the overlap rule except in the third position. The overlap therefore can be completely removed from B by removing the overlap in the third position. Rule B has a full set in the third position, symbolized by the entry "X," which is another way of symbolizing "don't-care." We can therefore remove the overlap by rewriting the third position of B so that it includes all the elements of the full set except the ones that appear in the third position of the overlap rule. This is done by taking the complement of (2, 3), the set in the third position of the overlap. Since the full set for the third row is the set of values (1, 2, 3, 4), the complement of (2, 3), is (1, 4)–that is, all the members of the full set except (2, 3). The result, B′, contains four rules and does not overlap rule A. A and B′ together specify the sixteen distinct rules that were covered by the original pair of composite rules, A and B.

Note that the condition that two composite rules not overlap (that is, that they be *disjoint*) is that their overlap rule must contain at least one condition square with an empty subset. The overlap rule in figure 6.2b has an empty subset in the third position.

In this example, removing the overlap was simple because B differed from the overlap rule in only one position. When the composite rule that we want to modify differs from the overlap rule in more than one position, the procedure is a little more complicated.

In figure 6.3, we start with the same rules, A and B, as in figure 6.2, but we remove the overlap from rule A instead of from rule B. This process is not as simple as the first one was. We replaced B by a single rule, B′. We find that we cannot replace A by a single composite rule; instead, we have to replace A by a pair of composite rules, A′ and A″.

This is because A differs from the overlap rule in two positions instead of one—the first and fifth positions, to be exact. Therefore, when we remove the overlap in the first position, as we do in figure 6.3b, we remove too many rules. We remove not only all the rules in the overlap set; we also remove rules that are not in the overlap set. These rules are the ones that overlap in the first position but do not overlap in the fifth. Therefore, we must restore these rules. We do this in figure 6.3c. There we calculate a second rule, A″, which overlaps rule B in the first position, but does not overlap in the fifth.

Let us discuss this in digital form, using complete sets of distinct rules.

The overlap rule contains the set of rules 11221, 11222, 11321, 11322.

When we rewrite A to get A′, we eliminate the rules 11221, 11222, 11223, 11321, 11322, 11323. The first, second, fourth, and fifth of these are the ones we want to eliminate. The third and sixth, however, are rules we want to retain.

We restore these rules by including a rule A″ that is equivalent to the two rules eliminated by A′. The digital form of A″ is 11(2, 3)23, which is shorthand for the two simple rules 11223 and 11323. These are the non-overlapped rules eliminated when we replaced A by A′.

If we follow this procedure, composite rules A and B are replaced by composite rules A′, A″, and B.

The extension of this idea to the case where the rule to be modified differs from the overlap rule in more than two positions is straightforward and easy to program.

D. THE NUMBER OF RULES IN AN "ELSE" RULE

Many decision tables include a rule called the "Else" rule. We encountered one in the answer to exercise 8 of chapter 2. An "Else" rule is a rule that applies when none of the others does. When we are concerned with the completeness of a table, we want to know how many rules are covered by the "Else" rule.

The straightforward way to determine this is illustrated in figure 6.4. We first calculate the number of rules in a complete table. We then calculate, for each rule other than "Else," the number of rules it covers. We add up the total number of rules and subtract the sum from the number of rules in a complete table. The difference is the number of rules covered by the "Else" rule. The calculations for a specific table are shown in figure 6.4a.

The analysis for more complicated cases should be done by a computer program. Without going into the details of such a program, let us consider briefly some of the characteristics of the digital codes we have been using that could be exploited by it.

Consider the first condition in the condition stub of figure 6.4. It has three values: 1, 2, 3. We know that we can describe a complete table as a set of four-digit condition codes and a fifth digit, which tells us what action code is associated with each condition code. By the rules of formation of the condition codes, we know that one-third of them will have the digit 1 in the first position; one-third will have the digit 2 in the first position; one-third will have the digit 3 in the first position. Since there are 36 rules in the complete table (see figure 6.4), we know that $36 \div 3 = 12$ rules will have a 1 in the first position; 12 will have a 2 in the first position; 12 will have a 3 in the first position.

When we examine rule 1 in figure 6.4, we see that it has a 1 in the first position. In chapter 5, section C, we learned how to calculate how many simple rules there are in a composite rule. In this case, the calculation tells us that there are $1 \times 2 \times 3 \times 2 = 12$ simple rules in composite rule 1. Since this is all the rules with a 1 in the first position that there are in a complete table, we know that we have accounted for all of them and that none can occur in the "Else" rule.

Rules 2 and 3 have a 2 in the first position. Each covers six rules; together they cover 12. Since they don't overlap, we know that no rule with a 2 in the first position can occur in the "Else" rule, since a complete table contains exactly 12 such rules and we have accounted for all 12.

Rules 4, 5, 6 have a 3 in the first position. Respectively, they cover 2, 2, and 1 simple rules; together they cover 5 simple rules. Since the complete table has 12 rules with a 3 in the first position, we now know that all the 7 rules covered by "Else" have a 3 in the leading position.

How many rules in a complete table have a 3 in the first position and a 1 in the second position? We can calculate this number in either of two ways. We can calculate the number of rules in composite rule 31XX–that is, a composite rule with a 3 in the first position, a 1 in the second position, and Don't-care entries in the third and fourth positions. This calculation tells us that there are 6 rules in composite rule 31XX.

There is an alternative way, which will be easier to calculate when the decision table is large. We know the total number of rules in a complete table—36 in our example. We have already determined that, since the first condition has three values, any one of these values will occur in one-third of 36, or 12, rules. How many of these twelve rules will have a 1 in the second position?

Again, we consult the condition stub and determine that the second condition has only two values. This means that one-half of all the rules in a complete table will have a 1 in the second position; the other half will have a 2.

The same reasoning applies when we are considering not a complete table but any composite rule with a Don't-care entry in the second position; half of the simple rules it covers will have a 1 in the second position, the other half will have a 2.

Therefore, of the 12 rules in composite rule 3XXX, 6 will be of the form 31XX and 6 of the form 32XX.

We calculate this number directly by dividing 36 by 3 and dividing the result by 2. Using fractions to represent the divide operation, we have $36 \times 1/3 \times 1/2 = 6$.

Rules 4, 5, and 6 cover 5 rules of the form 31XX. We saw above that there are 6 rules of this form. Since the missing rule does not occur elsewhere in the table, it must be included in the "Else" rule.

Similar reasoning tells us that all six rules of the form 32XX must be included in the "Else" rule. Together with the one rule of the form 31XX (which turns out to be 3132), we have accounted for all seven of the simple rules in "Else."

E. INCONSISTENCY: INTERCOLUMN AND INTRACOLUMN

A decision table can be inconsistent in two ways:

1. *Intercolumn* Two overlapping rules specify different action sets.
2. *Intracolumn* A condition-square entry contradicts another condition-square entry in the same rule.

We have already discussed how the first kind of inconsistency arises. An example is given in figure 6.5. The first and second rules overlap but they specify different action sets. Which invoicing procedure should we follow when, for example, we wish to bill a new customer for the purchase of a primary product for which no metallurgical specifications are on file and which is to be shipped rail-on-rail to the eastern part of the country? The code for the condition set specifying this procedure is 11221. This is covered by both the first and second rules, so we can't determine whether to follow invoicing procedure A or invoicing procedure B.

This kind of inconsistency can be readily detected by a program for analyzing and translating decision tables.

The second kind of inconsistency is harder to detect mechanically. It occurs when two or more rows in the condition stub are interdependent. The easiest kind of interdependency to illustrate is the kind that is shown in figure 6.6. In that figure, both the second and third rows have age as the fundamental common condition being tested. The difficulty arises because the age brackets that are significant for the second row (automobile insurance) are not the same as those significant for the third row (life insurance). This means some condition sets are not possible. In rule 4 of figure 6.6, the second-row entry specifies an age between 26 and 65, and the third row entry specifies an age between 16 and 25. Clearly, one person cannot validly be classified in both brackets.

This kind of interdependency can be a great deal more subtle than the obvious kind we have illustrated. This is because one row can be dependent on another in a completely arbitrary way. Suppose one of the conditions in the table was "Make of Car" and another was "Color of Car." In the early days of automobile manufacture, Henry Ford told his customers that they could have any color they wanted as long as they wanted black. In those days, a rule that specified *make* as "Ford" and *color* as "Seasick Green" would have been inconsistent, but the inconsistency would have been hard to detect mechanically

This latent or hidden interdependency between rows is very important. A good procedure designer should be aware of this and construct his table so as to detect the forbidden combinations and process them as coding errors. In practical problems, as we have noted, out of the large total number of possible rules only a few actually occur. A decision table should prescribe what to do for *all* the condition sets of a complete table. This includes those condition sets that should not occur. If they do occur, it is because an error in processing has been made, and this fact should, at the very least, be reported to users of the program's output.

Rule 1 of figure 6.6 illustrates some of the sloppiness that can creep into a procedure description if we disregard the kind of interdependency that we are calling *intracolumn* when it occurs in decision tables. The intent of rule 1 is clear: it says that people younger than sixteen are eligible only for life insurance; they will not be sold any kind of automobile insurance.

Unfortunately, rule 1, as it stands, specifies eight inconsistent simple rules. In fact, it is *not possible* to write a consistent rule that has a 1 in the third position. This is because of the interdependency of the second and third rows. None of the age brackets specified in the second row covers ages lower than sixteen. This means that none of them can be combined with the 0-15 category in the third row.

This example is clearly a contrived one. But it is not unrealistic. It is probable that most coding systems and programs now in use contain inconsistencies and incompatibilities of the kind we are discussing. They are overlooked because, as in the case of our example, it is obvious what the procedure is *supposed* to do. This blinds us to the fact that, as described, it doesn't do it. No harm is done until a coding error slips by the control clerk, and the program that should have been out first thing Friday limps painfully along (smearing several important files in the process) and finally goes down in flames at 2:30 Sunday morning. The operator hits the panic button; phone bells shrill in the night; and all hands are called to battle stations for the duration of yet another emergency.

F. THE COMPLICATED STRUCTURE OF SIMPLE PROCEDURES

In the preceding parts of this book, I have usually called myself "the writer" and called you "the reader." In part, this has been done in deference to the popular belief that a serious book must be impersonal and unreadable. In larger part, however, the avoidance of the first person singular has had more honorable motives; constant repetition of the words *I, me, my* can be both tiresome and vainglorious.

In this section I use a more personal form of expression, because I suspect that some of you may feel I have betrayed your trust. I started this book with a simple procedure for selling used cars of a rather unusual kind. Selling used cars is certainly a practical activity—as anyone with a television set can testify. Yet now I have gone from used cars to such abstract mathematical notions as vectors and sets and digital codes. Have I been guilty of a deceptive sales practice? Have I enticed you to read (perhaps even to buy) this book by luring you on with the promise of practicality and attempting to sell you impractical mathematical theory instead?

I don't think so. The minute we plan to mechanize procedures we find ourselves faced with the most difficult kind of mathematics, the mathematics of combinations—*combinatorial* mathematics. For example, one of the exercises at the end of this chapter is concerned with validating an order for shoes offered for sale in a mail-order catalog. Can anything be more practical? Can anything be simpler?

Yet when we examine this simple process in detail, we find that it is *not* simple. Although we are concerned with only one specific style of shoes, it comes in various colors, sizes, and widths. Not all combinations are stocked and the number of possible valid and invalid combinations is surprisingly large.

The example was suggested by the examination of a small portion of one page of a mail-order catalog of over a thousand pages. Most of the items described in the catalog are much more complicated and have many more features, options, colors, and so forth than shoes. In a practical sense, it is impossible to determine how many different valid and invalid orders can be placed for catalog items, particularly when we know that customers will make ingenious, unpredictable mistakes in providing order information. Given these difficulties, how can we devise a computer procedure to reject invalid orders; check valid orders against stock levels and fill them or back-order them as needed; accommodate returns, claims, adjustments, sales, seasonal offerings, additions to and deletions from the product line, and so on?

Let us consider another example, the invoicing procedure described in exercise 3.6. It was suggested by an actual procedure that was carefully studied by a team of people whose objective was to write a computer program for invoicing. The procedure in the example is much simpler than the real one. Even more important, the real procedure *did not exist before the study was undertaken.* This was because the owner, the general manager, the parts manager, and the pricing clerk all had a general idea of what the procedure was, but no two could agree on its details. It took several months of study by several people before a reasonable level of agreement could be achieved. Even then, the procedure was not completely specified, since some of the managers had the authority to alter its provisions if a particular circumstance warranted an exception to its rules.

Procedures like these seem simple only because we overlook our reliance on the judgment of people to make them work. People spot incongruities and impossibilities; they know what is happening around them and how it affects the response they are supposed to make in any given situation; when all else fails, they can make a phone call to find out how things really are as opposed to how they're supposed to be.

If we stop a computer procedure to place a phone call, we defeat the purpose of having a computer in the first place. In a few hours, a modern computer can easily duplicate the clerical efforts of a roomful of clerks working for several months. Can you imagine specifying procedures that are so precise and so detailed that the clerks never encounter a situation in which they have to ask a question or exercise initiative of any kind? Can you imagine them operating in a room without phones?

Consider yet a third example: the interior of the computer itself. Functionally, a computer is just an assemblage of interacting procedures that define the workings of adders, multipliers, shift registers, gates, instruction counters, location counters, device control commands, and so on. All these procedures are trivial compared with the simplest clerical procedure imaginable. What makes computer design difficult is not that internal computer procedures are complex, but that they are numerous, they have to be synchronized precisely, they have to be implemented at the lowest cost possible, they have to be executed very rapidly, *and they must never ask any questions.* A question in this case is a *machine check*—an indication that something has arisen that the computer cannot handle so that it has to stop until the problem either is solved by some outside agency or is bypassed—again by some outside agency.

If you are interested in seeing how the relatively trivial procedures taking place inside a computer are documented, take a look at the manuals provided for the use of the people who are supposed to maintain it. They are a vast, intricate tangle of symbols, lines, diagrams, explanatory paragraphs, pictures, and references to other works. They can be interpreted only with great difficulty and only by a highly trained specialist in the field.

Our interest is in procedures that are much more complicated than the procedures that govern how a computer computes. We cannot stop for unanswerable questions—the clerical equivalent of machine checks—if we ever expect to get anything done. Accordingly, we face much more difficult design and maintenance problems than the computer designer or field engineer. Our concern is with the programmer and systems analyst; the specialized tools we provide for them must be more powerful, flexible, and responsive to their real needs than the tools that describe how sensors, pulse generators, and switching gates are combined inside a computer.

But much of what has been done to provide specialized tools (Boolean algebra, Veitch-Karnaugh maps, etc.) for the computer designer suggests what should be done for the computer programmer or systems analyst. That is why we find ourselves considering set theory. We are concerned with developing a logical calculus. The logical tool we need is a practical extension of those that already exist. In the next chapter, we survey these briefly and consider how decision tables resemble them and differ from them.

REVIEW

Either redundancy or intercolumn inconsistency can occur when composite rules overlap. This is the kind of inconsistency a table has when the condition portions of two or more rules overlap and their action portions are different. It can easily be detected by a computer program.

When *redundancy* occurs, we calculate a composite rule—the *overlap* rule—which describes the set of overlapped simple rules. If the overlap rule is the same as one of the original rules, that rule can be dropped, since it is completely redundant.

When the overlap rule differs from both the original rules, we have a situation that complicates the procedure of checking a table for completeness. Completeness checking is done by counting the rules that appear in the condition entry and seeing if they add up to the number of rules in a complete table. Each composite rule is counted as if it were the set of simple rules it covers. This means that, when there is overlap, some simple rules are counted more than once.

Redundancy can be removed by replacing one of the original rules by an equivalent rule (or rules) that will cover only simple rules that are not overlapped. If the original rule and the overlap rule differ in only one square, this is easily done. The values in the overlap rule are removed from the condition square in the original rule. The resulting rule will cover all the original simple rules that are not overlapped. If the original rule differs from the overlap rule in more than one square, the procedure is a little more complicated, but the same principles for removing overlap apply.

Intracolumn inconsistency is a much more subtle, difficult, and important matter. It arises when not all combinations of values from two or more conditions are valid. It is the responsibility of the procedure designer to specify which combinations are valid and which invalid (possibly by using the "Else" rule as the general error-processing rule).

When an "Else" rule is included in a table, it includes all the rules that are not covered by the numbered rules. The structure of the "Else" rule itself can be analyzed by the use of the properties of the digital codes that describe the condition entry.

EXERCISES

1. a. Without expanding them, calculate overlap rules for the rules in exercise 4.7.
   b. Are any of the rules completely overlapped?

2. Remove the overlap in the composite rules of the previous exercise.

3. What rules are covered by the "Else" rule of the table given in answer to exercise 3.6?

4. In figure 6.5, calculate the overlap condition set. Modify rule 1 so that the overlap condition set is removed.

5. List all the impossible combinations of second- and third-row codes in figure 6.6.

6. In a mail-order catalog, shoes of different makes and models are identified by 10-digit codes made up of numbers and letters. The tenth digit is always the letter S, which identifies the product type as "shoe," thus differentiating it from the products in other sections of the catalog. The first seven digits identify the make and style of shoe. The eighth and ninth digits specify the color or color combination.

   For a particular style of shoe, the color codes (8th and 9th digits) are 61, 62, 51, 60, 54, 56, 55, 70 (given in the order in which they are listed in the catalog). Not all widths are available in these colors. Here is a table of available widths:

61: C, D, EE

62: D

51: C, D, EE

60: D

54: C, D, EE

56: B, C, D, E

55: B, C, D, E

70: B, C, D, E

Sizes range by halves from 7 to 12. Not all sizes are available in all widths. The table below tells which sizes are available for widths from B to EE:

B : 9 to 12

C : 9 to 11

D : 7½ to 12

E : 7 to 11

EE: 7½ to 11

Consider a table for checking the validity of an order for this style of shoe. It must check to see that the requested width is available in the requested color and the requested size is available in the requested width. Construct a decision table to reject the order if there is any inconsistency and accept it if there is not. Assume that the individual specifications of color, width, and size are selected only from these values:

Color: 61, 62, 51, 60, 54, 56, 55, 70

Width: B, C, D, E, EE

Size : 7, 7½, 8, 8½, 9, 9½, 10, 10½, 11, 11½, 12

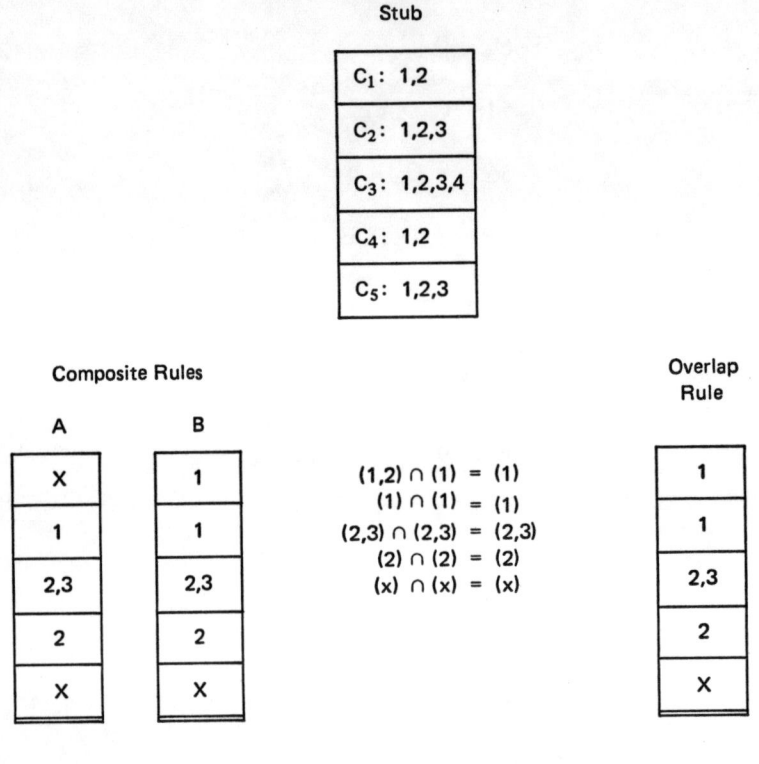

FIGURE 6.1 An example of complete overlap

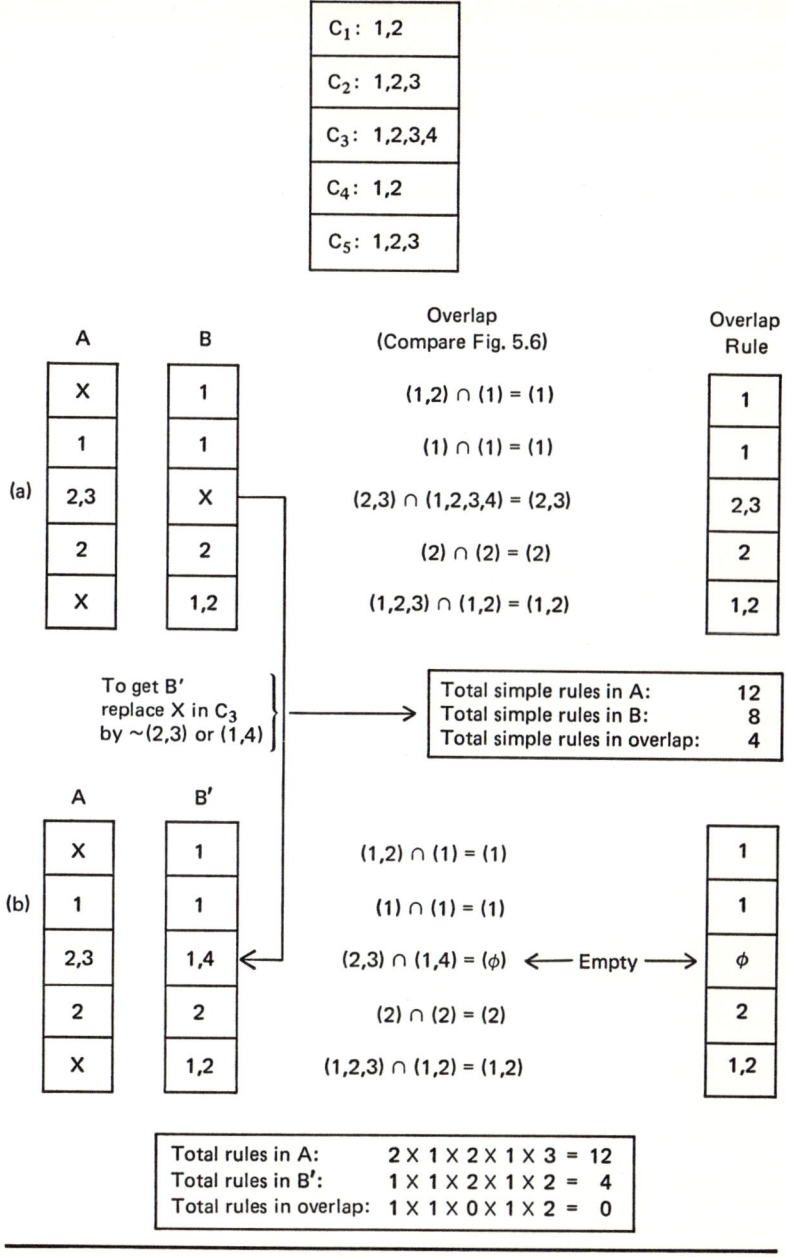

FIGURE 6.2 Removing overlap to get equivalent disjoint composite rules: I

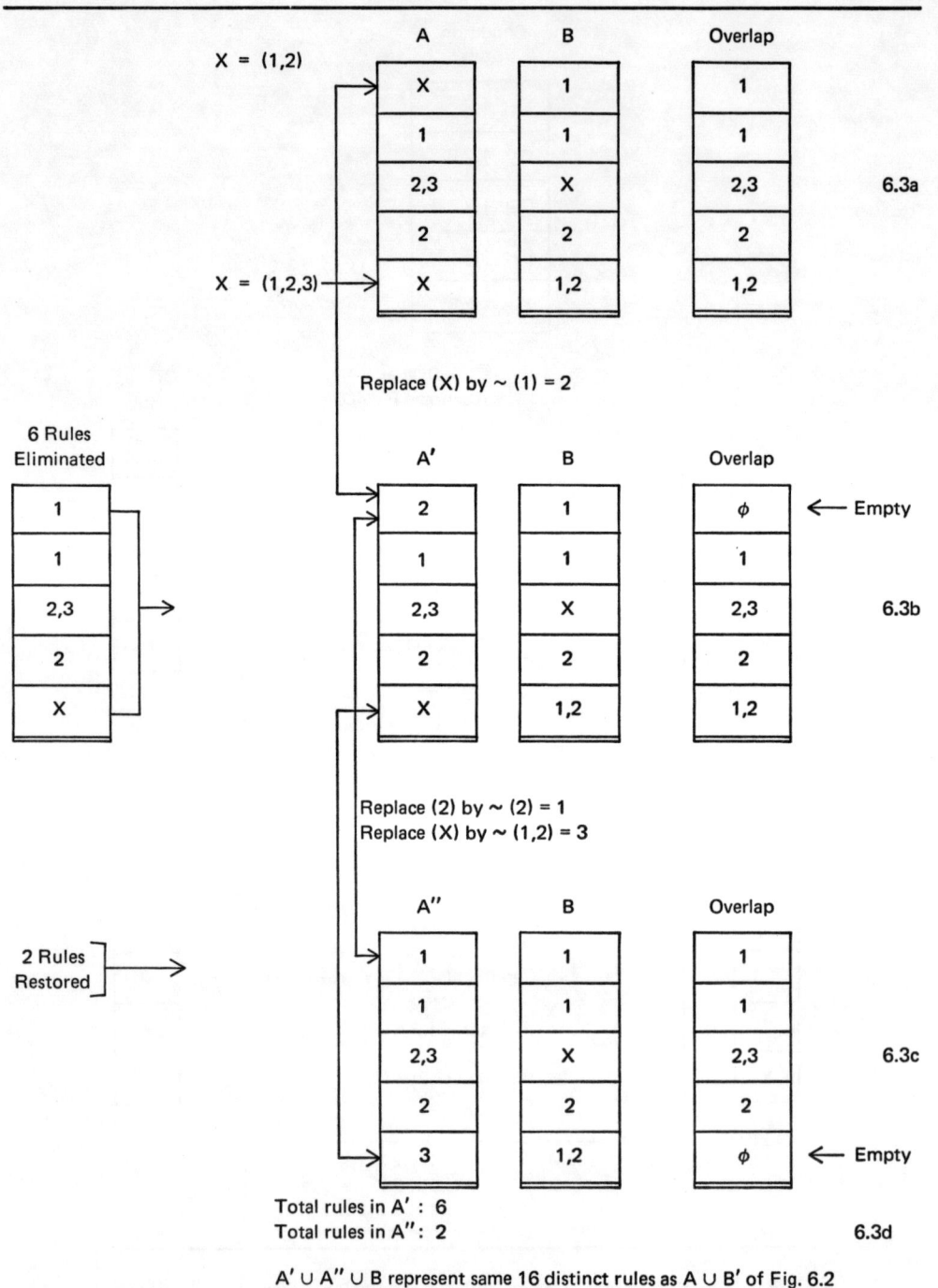

FIGURE 6.3 Removing overlap to get equivalent disjoint composite rules: II

(a)

		1	2	3	4	5	6	7
C_1 :	1,2,3	1	2	2	3	3	3	Else
C_2 :	1,2	X	1	2	1	1	1	
C_3 :	1,2,3	X	X	X	1	2	3	
C_4 :	1,2	X	X	X	X	X	1	
Actions		A	B	C	D	E	F	Error
Rule totals:		(12)	(6)	(6)	(2)	(2)	(1)	(?)

Total Rules in Table: 36
— Rules in rule 1:   −12
— Rules in rule 2:   − 6
— Rules in rule 3:   − 6
— Rules in rule 4:   − 2
— Rules in rule 5:   − 2
— Rule 6:   − 1

Else (remaining rules):   7

Further Analysis

    1XXX   covered by rule 1:   12
    2XXX   covered by rules 2 and 3: 12
    31XX   (except for 3132) is
                  covered by rules 4, 5, 6:   5

"Else" covers remaining rules
        3132 and 32XX: 7

(b)

Else		3	3
	≡	1	2
		3	X
		2	X
Error		Error	Error

FIGURE 6.4 Analyzing an "Else" rule

	1	2
NEW CUSTOMER? 1. Yes; 2. No	X	1
PRODUCT CLASS: 1. Primary; 2. Secondary; 3. Specialty	1	1
SHIP BY: 1. Ocean; 2. Rail-on-rail; 3. Rail-off-rail; 4. Truck	2,3	X
METALURGICAL SPECS ON FILE? 1. Yes; 2. No	2	2
AREA: 1. East; 2. Central; 3. West	X	1,2
INVOICING PROCEDURE	A	B

Overlap

FIGURE 6.5   Intercolumn inconsistency

	1	2	3	4	5
SEX: 1. Male; 2. Female	X	1	1	1	2
AUTO INSURANCE AGE BRACKETS: 1. 16–20; 2. 21–25; 3. 26–65; 4. over 65	X	1	2	3	1
LIFE INSURANCE AGE BRACKETS: 1. 0–15; 2. 16–25; 3. 26–45; 4. 46–55; 5. over 55	1	2	2	2	2
COMBINED AUTO-LIFE PLAN	None	A	B	C	A'
LIFE ONLY PLAN	X	Y	Z	U	V

Impossible Condition Sets

FIGURE 6.6   Intracolumn inconsistency

7

The Algebra and Geometry of Logic

A. LOGIC: SYMBOLS, DIAGRAMS, MACHINES

Until the middle of the nineteenth century, logic was a subject about which everything had been said by Aristotle over two thousand years ago. It was of very little practical use and was sterile and unproductive as a field either for research or for the development of new ideas. Kant, for example, could assert that "to all appearance, it may be considered as completed and perfected."

This situation was changed, as such situations frequently are, by the introduction of mathematical techniques: algebraic by Boole and geometric by Venn. Both had illustrious precursors like Leibniz, Euler, and DeMorgan, the latter contemporary with Boole. But the lines of development that have made logic a fertile field for both practical applications and academic research can fairly be said to have their origins in the publications of Boole and Venn.

Though our concern in this book is with the smallest, theoretically most trivial part of the vast structure of mathematical logic that has arisen since the publication of Boole's *Laws of Thought* in 1854, the discussion in this chapter is still much sketchier than the subject merits. The reader is urged to do outside reading. Inexpensive introductory books are currently available. One is by Martin Gardner,* whose ability to entertain while he instructs is the despair of lesser men. Another is by Lewis Carroll,† some of whose other writings may be familiar to the reader. Both

---

Logic Machines, Diagrams and Boolean Algebra, Martin Gardner (Dover, 1968).
†*Symbolic Logic* and *The Game of Logic*, Lewis Carroll (Dover, 1958).

books discuss the three interrelated ideas that are the subject of this chapter:

1. The use of letters and numbers in logical calculations
2. The use of diagrams in logical calculations
3. The use of machines or mechanical processes in logical calculations

A good way to start a discussion of how symbolic logic and decision tables are alike is to point out how they are different. They are different in their objectives. The objective of logic proper is to determine "truth" or "falsity"; the objective of a decision table is to determine a course of action. No matter how many logical variables are involved, the calculations of logic can lead to only one of two outcomes; the calculations governed by a decision table, on the other hand, can lead to as many outcomes as the number of rules in a complete table. The similarity between symbolic logic and decision tables is thus not in what they do but in how they do it. Both exploit symbolic and graphical techniques that enable the user to picture and manipulate complex interrelationships. And, most importantly, both the algebraic and geometric approaches are unified by the use of digital codes of the kind we have been studying.

Let us get down to cases by comparing a small decision table with a small truth table. Both are concerned with the truth or falsity of two propositions: (1) Tom is tall; (2) Dick is tall.

In logic, the word *proposition* is used for an assertion that can be either true or false. Not all sentences are propositions. Some are questions, some are commands, some are exclamations, some are nonsense, some are self-contradictory. For example, the sentence "The sentence you are reading is not true" is not a proposition, since it cannot meaningfully be said to be either true or false.

If we do not have some way of determining what we mean by "tall," the two sentences we are using in our example are not propositions either. A tall flyweight boxer might make a short basketball player. When we use a word of this kind in a proposition, we always imply a way to determine whether it does or does not apply.

To illustrate the use of the Tom and Dick propositions in a decision table, let us imagine the situation of a tall girl who would like to invite a partner to a Sadie Hawkins dance. She knows that Harry is tall but does not find him appealing and will invite him only as a last resort. Two of her friends have discussed Tom and Dick in glowing terms but have never discussed their heights. She can, however, call them and ask the questions we display in the condition stub of figure 7.1a. Thus her way of determining whether they are tall is to make two phone calls. The decision table displays the four courses of action that will be determined by the answers she gets.

Figure 7.1b displays a truth table in decision-table form. In the condition stub, we have our two propositions in the form of assertions instead of questions. In the action stub, we have a compound proposition: "Tom is tall and Dick is tall." The word *and* is a logical connective. The decision table of figure 7.1b defines what we mean by *and*, by telling us how the truth or falsity of the compound proposition "Tom is tall and Dick is tall" depends on the truth or falsity of the individual propositions of which it is composed. Specifically, it tells us that the only time the compound proposition is true is when both individual propositions are true. Otherwise, it is false.

Now what we have said about the two specific propositions "Tom is tall" and "Dick is tall" is true about *any* two propositions connected by *and*. To show that a truth table covers a general case rather than a specific instance, we employ the customary mathematical convention of using letters to represent propositions when we state general principles, conventions, or rules.

The letters A and B, used in the various truth tables of figure 7.2, stand for *any* propositions. The individual truth tables define the *logical connectives* or, more properly, *logical functions: and, or, not,* and *implies*.

It is important to note how the use of these words to denote logical functions compares with their use in everyday speech. The resemblance is closest for *and* and *not*. It starts to depart from English usage with *or*. The full meaning of *or* as it is used in English depends upon the context in which it arises. An emphatic "It has to be either this *or* that" usually means that it has to be one or the other but it cannot be both. This is a logical function, too—called *exclusive or*—but it is not the one whose truth table appears in figure 7.2b. That one is *inclusive or:* one or the other *or both*.

The departure from English meaning is greatest for the logical function *implies*. The ordinary English meaning usually expresses the idea of something like a causal connection between the antecedent and consequent. This is surprisingly hard to formalize. The implication pictured in figure 7.2d is not of this kind. It is called "material implication" and merely defines "A implies B" as a logical function that is true whenever either A is false or B is true. The consequences are surprising and have been the subject of much mystification. Two of them in particular—"A false proposition implies *any* proposition" and "A true proposition is implied by *any* proposition"—have caused much soul-searching among people who expected more from logic than *that*.

The important lesson to be learned from the difference between words as English and words as the names of logical functions is the one we stressed earlier: the distinction between *meaning* and *structure*. It is unfortunate that words like true and false continue to be used about logical functions, since they obscure the fact that all we are talking about is entities, objects, *variables* whose permissible range is restricted to two distinct values—usually represented by 0 or 1, though T and F, *On* and *Off, North* and *South*, or any other representations of two different states can also be used.

In terms of decision tables, the definition of truth tables is particularly simple. Truth tables are limited-entry decision tables that always have one action row and as many condition rows as there are *input* or *argument* variables. In figure 7.2, the *a, b* and *d* tables are tables for logical functions of two arguments—that is, they have two inputs and one output. Accordingly, they have two condition rows and one action row. The *c* table is a function of one argument, so it has only one condition row. The only permissible entries anywhere in the table are (0, 1) or any other two distinguishable symbols.

This tells us a surprising amount. Consider the action row, for instance. How many action squares will it have? Clearly, it will have as many as the number of rules in a complete table, since it must specify a 0 or a 1 for every possible combination of conditions. Now take a look at the actual entries along any action row. What do they look like? They look very much like a digital code. That is just what they are. In fact, if we generate the condition combinations in the way described in chapter 5, *we can define any logical function as the sequence of zeroes and ones that goes from left to right along the action row.*

Consider the two-variable tables of figure 7.2. Nothing significant would change if we replaced A by Q and B by Z everywhere they occurred in the table. Therefore, they give no information except the information that we are dealing with two logical variables instead of one, or five, or some other number. In the tables, the condition sets are always presented in the order 00, 01, 10, 11. We can standardize on this sequence, since it is as good as any other. This means that the only parts of the table that are fundamentally informative are the symbols that represent the logical functions $\wedge, \vee, \supset$ and the (0, 1) pattern in the action row. Therefore, it is valid shorthand to define two-variable logic functions as we do below:

Word	Symbol	Code
AND	$\wedge$	0001
OR	$\vee$	0111
IMPLIES	$\supset$	1101

The fact that we can do this opens up all sorts of possibilities that were not apparent before we established the one-to-one correspondence between four-digit binary numbers and logical functions of two variables. One of the more obvious is that it is now possible to state *exactly how many two-variable logical functions there are.* These are exactly as many as there are four-digit binary numbers: sixteen.

The reader should investigate the sixteen possible logical functions for himself. What does he make of combinations like 0000 or 1111, which are, respectively, False and True, no matter what the input values are? Do the patterns 0011 and 0101 tell him anything? Does the first one tell him, for example, that A is the only variable that counts and he can forget about B? Does the second one give him the same kind of information about B? Can he think of a name for the pattern 0110?

These investigations are not suggested capriciously. A major objective of traditional logic is to examine a complicated logical expression and determine what entries to make in the action row. These entries then define the logical expression as a particular logical function. If all the action entries turn out to be zero, the function is *identically false*—that is, it contains a contradiction. If all the entries on the action row turn out to be one, the function is *identically true*—that is, it is a tautology. If the action-row entries are neither all zero nor all one, the logician's interest then shifts. Since his traditional interest is in the True entries, he selects the rules that have a 1 in the action row and attempts to combine them into as few composite rules as possible. He does this in order to get a simpler logical expression than the (perhaps) complicated one he started out with. If he gets a 0011 pattern in the action row, for example, he knows that his expression is merely A (or whatever other symbol he used in the first row of his table). If he gets 0101, on the other hand, he knows that A is irrelevant and his resulting expression is merely B. If he gets 1100, he knows that his expression is ~A or $\bar{A}$ or A', or whatever other expression he uses to symbolize "not A." Everything he wants to do is readily expressible as some form of the combining and expanding procedures we discussed in the earlier parts of this book.

These are also of interest to the electrical engineer who wishes to design the least costly digital circuit that will generate a given pattern along the action row. He may describe his work by using words like *true* and *false*, but he doesn't really mean it. He really means "Pulse" when he says "True," and "No Pulse" when he says "False." We can think of his attempt to design a digital circuit for a particular logical function as an attempt to develop a "minimum-cost" set of composite rules to cover the simple rules that have a 1 in the action row.

Figure 7.3 illustrates a comparison between a three-variable decision table and a three-variable truth table. The condition portion of both tables is still limited-entry, but in the decision table we introduce an extended-entry action row. This highlights the difference between the decision table and the truth table. In the truth table, we are always restricted to (0, 1) entries along the action row. In the decision table, we need observe no such restriction.

Let us conclude by stressing once again the fundamental importance of digital codes when we talk about logical processes. Our concern is with "mechanizing" logic, in the broadened sense the word *logic* has come to acquire in recent years. The three techniques we discuss in this chapter (maps, symbols, codes) are interrelated and can be translated back and forth readily. But if we plan to mechanize logic by using a digital computer, the fundamental technique is the technique employing digital codes.

B. VENN DIAGRAMS: LETTERS AND CIRCLES

In previous chapters, we talked about sets. Earlier in this chapter, we talked about propositions. For certain purposes, we can convert one idea to the other. For example, when we say "Tom is tall," we can understand that to mean "Tom is a member of the set of tall people." The easiest approach to an understanding of Venn diagrams is to think primarily in terms of sets and to interpret propositions as asserting or denying the inclusion of a specified set of objects in a set defined by possession of a particular attribute.

For example, in figure 7.3a we displayed a decision table depicting the order of preference of a girl who was seeking a tall, dark handsome date but was realistic enough to consider settling for something less if she had to.

In figure 7.4a, we denote "Tall," "Dark," "Handsome" sets by circles with identifying labels. These circles are enclosed in a rectangle, which represents the "universe" for our present purposes. In this universe, the word *tall* means the same as (Ernie, Frank, George, Harry). *Dark* is the same as (Chuck, Dave, George, Harry). *Handsome* is the same as (Bob, Dave, Frank, Harry). *Not tall* and *not dark* and *not handsome* is the same as (Andy), the desperate last resort of Ms. 7.3a.

Figure 7.4a is a Venn diagram. It uses geometric techniques to explain logical relationships.

In this particular instance, what does the Venn diagram do? For one thing, it makes it clear which men are members of two or more sets. This can, of course, be determined by examining the lists of set members, but that takes a good deal more work than finding where the appropriate circles overlap in figure 7.4a.

It also makes vivid the fact that we are dealing with a finite universe. Whatever the words *tall, dark, handsome* mean outside rectangle 7.4a, whatever connotation they may have in terms of ideas about height, or complexion, or facial contours, those meanings and connotations are completely irrelevant. Within the confines of 7.4a a man is tall if he is Ernie, Frank, George, or Harry; he is dark if he is Chuck, Dave, George or Harry; he is handsome if he is Bob, Dave, Frank or Harry. Opinions or measurements have nothing to do with determining how the words are to be applied.

By setting up complementary sets to the sets of figure 7.4a, we turn its universe inside out, so to speak. Calling the complementary sets "Short," "Light," "Ugly," we get the Venn diagram of figure 7.4b with Andy snuggled happily at its center and Harry, the central figure of 7.4a, wandering desolately around its outer fringes. Tastes change; perhaps Andy is the ideal man of the sixties and seventies, whereas Harry represents the ideal man of earlier days.

What we are doing is illustrating the kinds of quick visual insight that pictures, graphs, and other geometric displays can provide. Let us examine

Venn diagrams to see how they provided insight in resolving the kind of question that originally motivated their development.

Classical logic expressed chains of reasoning as long strings of syllogisms called *sorites*. The syllogism was the fundamental unit. It was categorized by figure and mood. Students of logic used to memorize all the "valid" combinations of mood and figure. As an aid to memory, they used mnemonic verses that incorporated names like Barbara, Darii, Celarent, and Ferio. The vowels in the names described the "quantity" and "quality" of the premises in the valid forms of syllogism.

A typical syllogism is the one illustrated in figure 7.5. In verbal form, it goes like this:

>All tall men are handsome.
>Some dark men are tall.
>Therefore, some dark men are handsome.

There are three sets involved, which, in a Venn diagram, means three circles. The first two statements in the syllogism are called premises. They make assertions about set membership. The last statement is called the conclusion. It is also a statement about set membership.

What does the first statement assert? In logic, it is interpreted to mean, "If a man is tall, he is also handsome." This does not assert that there are, in fact, tall men. What it does assert is that there are no men who are both tall and ugly. If we look at figure 7.5a, we find that the circle representing the set of tall men is colored black everywhere except where it overlaps the handsome circle. This is the way we indicate the meaning of the first premise. Black means that a set or subset has no members. The effect of coloring part of the tall circle black is to indicate that *if* the tall circle has any members they are to be found only in that part of the tall circle that overlaps the handsome circle.

Unlike the first premise, the second premise, "Some dark men are tall," is usually interpreted as "There *are* dark men who are tall." In other words, it asserts that some members of a particular subset actually exist, whereas the first premise merely tells us that a particular subset has no members—that is, its putative members do *not* exist.

In figure 7.5b, we diagram the second premise by shading in the dark-tall subset. The shading indicates that the shaded subset has members.

In figure 7.5c, we combine 7.5a and 7.5b. In the upper half of the tall-dark subset, the shading is eliminated by the solid black of the tall-ugly subset. This overrides the effect of the shading. The part left shaded is therefore the only part of the tall-dark subset that has members. In our example, this turns out to be the tall-dark-handsome subset and the conclusion is justified, since, in terms of our assumptions, there do in fact exist dark men who are handsome.

Both the capabilities and limitations of the purely geometric approach can perhaps be guessed at this point. Drawing intersecting circles helps us picture the detailed workings of set-subset relationships. But it is not a really practical form of calculation, particularly if we have many sets. We need the kind of help that synthetic geometry got from algebra. We need a kind of "analytic geometry" of Venn diagrams. This exists. It is usually called Boolean algebra, after George Boole, who first described it, though its form was later modified by the work of others.

So far, we have applied descriptive labels to identify the circles we have drawn. This was done to make the examples concrete rather than abstract. But the syllogism whose truth we established would be valid for *any* three subsets that were related to each other in the same way. In other words, what we proved for *tall*, *dark*, *handsome* circles is also true for any circles A, B, C whose properties are described by the syllogism:

>All A is C.
>Some B is A.
>Therefore, some B is C.

To prove general theorems by means of Venn diagrams, we ordinarily use letters instead of words to identify the circles. This is what we shall do in much of the next section, in which we discuss Boolean algebra. After all, algebra is little more than a set of rules for juggling letters and other symbols.

C. BOOLEAN ALGEBRA: LETTERS AND NUMBERS

In Boolean algebra, *and* ($\wedge$) and *or* ($\vee$) are treated in much the same way as *times* ($\times$) and *plus* ($+$) are treated in ordinary algebra. Multiplying, factoring, and combining symbolic expressions also follows many of the same rules. In performing these algebraic operations, it is inconvenient to have a special symbol for *not* ($\sim$) preceding the letter identifying the set it complements, so an overbar ($\sim A \equiv \bar{A}$) or a prime symbol ($\sim A \equiv A'$) is more commonly used.

An ordinary algebraic expression like $A(B + C)$ is a factored form of $AB + AC$, which in turn is really shorthand for $(A \times B) + (A \times C)$. We don't see the times signs because we let the times operation be signified by mere juxtaposition instead of by an explicit symbol; that is, we write AB when we mean $A \times B$.

In figure 7.3b, a logical expression, $A \wedge (B \wedge \sim C)$, is expressed as a logical function. The rules that have a 1 in the action square are 5, 7, and 8. The respective condition sets are 100, 110, and 111. Using the expertise we acquired in chapter 4, we combine the first two rules to get composite rule 1X0, and combine the second and third to get 11X. What do these two composite rules tell us?

1X0 tells us that the composite statement $A \wedge (B \wedge \sim C)$ is true whenever A is true and C is not true ($A\bar{C}$). 11X tells us that the composite statement is true whenever both A and B are true (AB). We don't know why this should be so, since we don't know what principles were used to determine where to place zeroes and ones in the action entry. But we do know that, for the given pattern of zeroes and ones, the two composite rules describe which combinations of A, B, C truth values give us the "true" symbol 1.

We use the operations of Boolean algebra to derive the same result by symbolic manipulation. First, we rewrite A ∧ (B ∧ ∼C) as A(B + $\bar{C}$). In the latter expression the ∧ symbol is implied by placing A next to the left parenthesis, the way we do for × in ordinary algebra. The ∨ symbol is replaced by a +. The ∼ is replaced by an overbar. In Boolean algebra, A(B + $\bar{C}$) is a factored form of AB + A$\bar{C}$. The rule is the same as that for ordinary algebra:

$$A(B+\bar{C}) = AB+A\bar{C}$$

The expression $AB + A\bar{C}$ gives us exactly the same information as our two composite rules. AB represents the intersection or overlap of the A and B circles in a Venn diagram; it describes the subset for which both A and B are true. A$\bar{C}$ represents the intersection or overlap of A with all the space in the Venn rectangle that is *not* contained in circle C. The effect is to give us that part of A which is not included in C, the part in which A is true and C is not true. At this point, we have three different but equivalent ways to picture set-subset relationships:

1. *Intersecting circles* Any circle divides the universe (the rectangle within which it is inscribed) into two parts. When it intersects another circle, it divides that into two parts also (and, of course, is itself divided into two parts by the other circle). The part where it overlaps the other circle represents the subset that is common to both. A third circle divides both the universe, the original circles, and their common subset into two parts. And so on, each additional circle dividing the universe and all preexisting sets and subsets into two parts. (Reexamine figure 7.4 to study an example of this process.)
2. *Letters* Each letter represents a set. The part of the universe not included in the set is represented by the set identification letter with an overbar or a prime symbol. The subset common to two sets is represented by the two set identification letters written side by side. The set consisting of all the members of both sets is represented by the set identification letters with a + between them. And so on, for more than two sets.
3. *Digital Codes* The code has as many digits as the number of sets being considered. The sets are arranged in order so that the first digit represents the first set, the second represents the second set, and so on.

The difference between the letter representation of sets and the digital-code representation is illustrative of the difference between expressing information by form and expressing it by position. The intersection or overlap of the sets A and $\bar{C}$ can be represented either as $A\bar{C}$ or $\bar{C}A$. The order in which the letters appear is not significant. In the coded representation, on the other hand, the first position is reserved for A, the second for B, and the third for C. Thus only one coded representation is possible, 1X0, the code which represents the A value as 1, or true; the C value as 0, or false; and the B value as X, or don't-care, or irrelevant, or either true or false.

Our purpose in using Venn diagrams is not to prove syllogisms. Rather it is to map a statement of policy—a description of alternative courses of action and the specific circumstances that determine which course is to be followed. This is illustrated in figure 7.6. Figure 7.6a maps the strategy described in decision table 7.3a. Figure 7.6b maps the strategy described in truth table 7.3b.

The relative preferences of the girl seeking the tall-dark-handsome companion are entered in the subsets created by the overlapping of the tall-dark-handsome circles. The circles have been renamed A, B, C to permit the entry of information describing the subsets. This is done in two forms: (1) letters; (2) digital codes.

Since there are three circles in this universe, the smallest subsets are all described by combinations of three letters: A, B, C. They are distinguished from one another by the placement of overbars over individual letters; the presence or absence of these indicates whether the subset excludes or includes members from the set in question.

All the digital codes describing the smallest subsets correspond to simple rules—that is, they do not have don't-care entries. A 0 in the code corresponds to an overbar in the letter identification; a 1 corresponds to the absence of an overbar. To illustrate this correspondence, the letters have all been written in ABC order. It is important to remember, however, that the order of the letters is as irrelevant in Boolean algebra as in ordinary algebra; $A\bar{B}\bar{C}$, $A\bar{C}\bar{B}$, $\bar{B}A\bar{C}$, $\bar{B}\bar{C}A$, $\bar{C}A\bar{B}$, $\bar{C}\bar{B}A$ all denote the same subset, the one represented by code 100, the subset of A which is common to neither B nor C.

Throughout the preceding discussion, we have used the term *smallest subset* to identify the subsets that represent the finest subdivisions of the universe, the subsets that are not themselves further subdivided. In a three-circle diagram, the subset AB is divided into two parts by circle C: (1) ABC, common to all three circles; (2) $AB\bar{C}$, common to A and B, but not common to C. AB is therefore not a "smallest" subset, since it is composed of two "smaller" subsets.

Using words like *small, smaller, smallest* in this way can be misleading, since they might be taken to refer to the number of members in a set. To avoid the possibility of this kind of misunderstanding a word has been coined to denote what we have been calling a smallest subset. The word is *minterm*. *Minterms* correspond to simple rules. For three sets A, B, C, there are the same number of minterms as the number of rules in a complete limited-entry table with three condition rows. Their letter designations are $\bar{A}\bar{B}\bar{C}$, $\bar{A}\bar{B}C$, $\bar{A}B\bar{C}$, $\bar{A}BC$, $A\bar{B}\bar{C}$, $A\bar{B}C$, $AB\bar{C}$, ABC. Their code designations are 000, 001, 010, 011, 100, 101, 110, 111.

Figure 7.7 summarizes most of the conventions and relationships we have discussed in this chapter. The analogous symbols of set theory and Boolean algebra are described, listed, and compared in 7.7a. Set-subset relations for three sets are illustrated in 7.7b, in which the conventions of Boolean algebra are paralleled by corresponding decision-table conventions.

Our purpose in studying topics like set theory and Boolean algebra is not to acquire an understanding of the subjects themselves. Rather it is to get a feel for "logical structure." Much of this feel is imparted by logical maps. The logical maps in current use are most commonly coded, rearranged Venn diagrams. This means they are "two-valued" maps—that is, both inputs and outputs are restricted to the range (0, 1). In decision tables, neither the inputs (*arguments*) nor outputs (*function values*) need be restricted to two values. Let us see how we can extend the ideas of two-valued maps to more general mapping techniques.

D. TWO-VALUED LOGICAL MAPS

If we use circles to denote sets, we cannot draw a Venn diagram that shows all possible subsets for four or more sets.

We can see why by examining the progression of illustrations in figure 7.8: 7.8a shows the universe divided into two parts by a single set; 7.8b shows two sets and four parts; 7.8c shows three sets and eight parts; 7.8d shows four sets but—alas!—only fourteen parts instead of sixteen. We cannot draw circles in such a way as to make two circles across the way from each other overlap without overlapping at least one other circle.

Various techniques, some suggested by Venn himself, even better ones by Lewis Carroll, have been suggested to overcome the difficulty. Ellipses and doughnut-shaped sets have been among the suggested expedients. None is really satisfactory. The objective of using the diagrams for acquiring insight is defeated by their complexity when more than three variables are involved.

What was the purpose served by the diagrams in the first place? In effect, it was to enclose within one circle all the subsets that make up the set represented by the circle. This purpose was actually only realized for the positive sets, A, B, C, ... The negative sets, $\bar{A}, \bar{B}, \bar{C}, \ldots$ are not represented by subsets confined within circles; instead they are represented by subsets found in the area of the rectangle outside a given circle. Except for the logician's preoccupation with something he calls "truth," there is no valid reason for slighting the negative or complementary sets. We saw this in figure 7.4, where we shifted the center of attention from Hero Harry to Anti-hero Andy.

The Veitch-Karnaugh map is an extension of the Venn diagram in which all the subsets making up a given set are, in a generalized sense, contiguous.

The structure of a four-variable Veitch-Karnaugh map is dissected and then combined in figure 7.9.

Four two-valued variables divide the universe into sixteen parts. The eventual map thus consists of a square, representing *all* the subsets that make up the universe.

The subsets themselves are grouped. Those in the first two columns are the subsets that make up A'; those in the last two columns make up A. The subsets in columns 1 and 4 make up B'; those in 2 and 3 make up B. We can think of the columns making up B' as contiguous, since we could make them so by shaping the square into a cylinder with a vertical axis.

C', C, D', D are treated similarly, except that they are contiguous along rows instead of columns—that is, horizontally instead of vertically. To make rows 1 and 4 contiguous, we roll the square into a cylinder along its horizontal axis.

The *a, b, c,* and *d* sections of figure 7.9 show how the square is divided into positive and negative halves by sets A, B, C, D. Superimposing these four on the same square produces the map itself.

E. INTEGER-VALUED POLICY MAPS

How does a designer of digital circuits use a Veitch-Karnaugh map in his work? He gets a set of specifications that tell him when to produce a pulse (1) and when not (0). These specifications can come in many ways. All of them tell him, in one form or another, what inputs he is to use and what combinations of those inputs are to produce a pulse.

He may, for instance, get the description given in the legend of figure 7.10, the Boolean function $AB'C'D' + B'C + A'B'C' + AC'D + A'BC'D$. Each of the terms in this expression tells him where to make "True" or "1" entries in the Veitch-Karnaugh map. The first, $AB'C'D'$, for example, corresponds to code 1000, and accordingly he puts a 1 in the upper right corner of the map. The second, $B'C$, corresponds to code X01X, which in turn is shorthand for the four simple codes (or simple rules) 0010, 0011, 1010, 1011. Accordingly, he puts a 1 in each of those squares. $A'B'C'$ gives 0000 and 0001; $AC'D$ gives 1001 and 1101; $A'BC'D$ gives 0101. The resulting pattern is shown in figure 7.10.

By inspection, the designer can see that all the B' squares have 1 entries. He can therefore account for eight of the ten entries in the square by running the B input through an inverter, an electrical implementation of the logical function *not*. An inverter produces a 1 when the input has a 0, and a 0 when the input has a 1. Now only two of the original ten squares (0101, 1101) require attention.

These squares are the middle part of the $C'D$ row. The other two squares in the row also have 1 entries. This means that a circuit which *ands* C' and D will generate all four. The first and last will also be generated by B', but this does no harm, since the *or* function he uses is *inclusive or*, the one whose truth table we gave in figure 7.2b. He runs the C input through an inverter, *ands* it with D, and *ors* the result with B'.

The result is a little surprising. It is not only much simpler than the original expression, it also treats input A as irrelevant. Why? The question is left as an exercise for the reader.

In generalizing the logical map for our purposes, we find at once that we have a much more difficult problem. We can no longer rely on (0, 1) inputs and outputs. We can always code our inputs and outputs as integers, but it would be extremely inefficient to restrict ourselves to only two values.

What we want is therefore not a *logical* or (0, 1) map; we want what might be called an integer-valued *policy* map—a map which describes a correspondence between a vector of condition values and a vector of action values. In other words, we want a map that tells us what to do when we are confronted with a particular set of circumstances.

Figure 7.11 shows a simple policy map for the vintage-car decision table of figure 2.1. The vertical axis of the map is the "Make" dimension; the horizontal axis is the "Condition" dimension. The result is nine "situation" boxes. In each of these, we indicate a course of action by placing one of the action codes we described in figure 4.1.

What have we gained by doing this? In the circuit design illustration, we learned that A was a completely irrelevant input. Our achievements in this simple case are not quite as spectacular. We have learned something, however. For one thing, if the make is Duesenberg, the car's condition is irrelevant; the decision will be the same no matter what condition the car is in. If the make is *not* Duesenberg and the car is either running well or not running, then the make is irrelevant. For those two values of "condition," the action is the same whether the car is a Cord or a Reo. In fact, we do not have to determine the make of the car unless it is not a Duesenberg and is running poorly.

A policy map permits us to develop simple sets of composite rules in much the same way as a logical map does. In a logical map, the expression B' + C'D, expressed as a pair of composite rules, becomes X0XX and XX01. What about a policy map?

The policy-map problem is a good deal more complicated, since we have inputs and outputs with more than two states, but the simplifying and interpreting procedures—"consolidating" and "expanding" (chapter 4, sections C and D)—are fundamentally the same.

In comparing logical maps and policy maps as techniques for developing insight, two points should be remembered:

1. Even for the relatively simple purposes for which they are intended, logical maps are of limited utility.

2. For practical problems, the "insight" to which we should appeal is not that exercised by a human being inspecting a two-dimensional or three-dimensional diagram. It is, instead, that exercised by a human being in writing a computer program that will "inspect" an array of many dimensions.

To illustrate the limitations of logical or policy maps, let us consider the Veitch-Karnaugh map in figure 7.10. It has only four variables. What does such a map look like for five variables? Six? Seven? More?

For large numbers of variables, maps get very complicated indeed. Thus, as the need for insight increases, the map's ability to provide it decreases. Then, too, the insight is most easily expressed as an understanding of logical implementations that use only the functions *not, and,* and *or.* As we have seen, these are not the only ones possible. *Nand* and *nor,* for example, are also commonly used, since electronic circuits to realize them are simple and inexpensive. But the more ways we devise to generate the 1 entries in a map, the harder it becomes to see at a glance which is best, particularly since "best" can mean a combination of least cost, most reliable, and other practical considerations.

The result is that, for practical problems, "insight" is expressed by means of a computer program. Such a program converts a general-purpose digital computer into a "logic machine." The insights developed by a human being for the simple cases that can be described by maps are expressed as algorithms that the computer can apply to cases for which maps would be impractical.

Later we discuss briefly how this might be done. Before we do this, let us consider figure 7.12, a policy map for the invoicing procedure described in exercise 3.6.

Figure 7.12 has two parts. Figure 7.12a displays three-digit action codes placed in 68 out of 72 possible "situation" boxes. Each situation box corresponds to a simple rule based on a condition stub with 6 customer-class, 4 amount-class, and 3 product-class conditions.

Figure 7.12b represents the "inverse" of figure 7.12a. In this figure, situation codes are placed in action boxes that are defined by 7 discount codes, 2 consignment codes, and 4 codes for terms of payment.

Figure 7.12a tells us what course of action to take for any given set of conditions; figure 7.12b tells us which courses of action are actually followed and under what different conditions any particular course of action is the one to be followed.

The information in figure 7.12 is exactly the same as that given in exercise 3.6. The difference in the two is in the relative ease with which questions about the invoicing procedure can be answered by consulting the map rather than the verbal description. For example, an examination of figure 7.12a tells us immediately where we have failed to specify courses of action for certain combinations of conditions. These are simple rules 311, 321, 532, 542. These constitute the "Else" rule for the table.

F. USING COMPUTERS TO ANALYZE LOGICAL MAPS

What does a logical map or its generalization, a *policy* map, look like inside a computer? This, of course, depends upon the programmer, or systems analyst, or both.

Figure 7.13 shows a typical computer realization of the Veitch-Karnaugh map of figure 7.10. Figure 7.14 does the same for the invoicing-procedure policy map of figure 7.12.

Figure 7.13a displays a correspondence between argument or "input" sets of four binary variables and function or "output" values of one binary variable.

How does a computer "examine" this representation of a logical map to get the kind of insight that figure 7.10 gave us when we inspected it visually?

Each four-digit input in figure 7.13a corresponds to what we have previously called *smallest subsets* or *minterms*. Larger subsets are describable as collections of these minterms. The set A, for example, is the collection of eight four-bit inputs whose first digit is a 1; the set A' is the collection of eight four-bit inputs whose first digit is a 0. The set D is the collection of eight four-bit inputs whose last digit is a 1. The set C'D is the collection of four four-bit inputs whose third digit is 0 and whose fourth digit is 1.

The general rule is as follows: a set represented by a single letter is a collection of eight inputs; a set represented by two letters is a collection of four inputs; a set represented by three letters is a collection of two inputs; a set represented by four letters is a minterm, one input.

It is this fundamental rule that guides our algorithmic search for insight. In analyzing figure 7.13a, our objective is to find a simple way to describe the ten inputs that lead to a 1 output.

These ten are displayed in figure 7.13b. Underneath the inputs themselves are two rows giving counts of the number of zeroes and ones in each column. What information can we derive from these counts?

For one thing, we can determine immediately that the set B' covers eight of the ten rules in figure 7.13b. We determine this from the count of zeroes in the second column: 8. As we noted above, a collection of eight distinct inputs with a fixed value in one position makes up a single-letter set. In this case, the fixed value is a 0 in the second or B position. The remaining positions run through all possible combinations of (0,1) values.

If we subtract the contributions of the B' collection of inputs from the column sums, we get the "reduced" sums shown in figure 7.13c. These sums tell us that, in the two remaining rules, one rule has a zero in the first position, and one rule has a one. Both have a one in the second position, a zero in the third position, and a one in the fourth position. This is one way to describe composite rule X101, set BC'D. We already know that rule X001 (set B'C'D) is included in the B' set. Therefore, the two remaining rules can be consolidated with their B' counterparts to give us XX01, set C'D.

The preceding paragraphs constitute a verbal description of part of an algorithm for processing Veitch-Karnaugh maps by computer. Our interest in such algorithms is not in their details. Rather, it is in how sequences of digital operations achieve the same results as visual inspections of Venn diagrams, Veitch-Karnaugh maps, and "policy" maps—computer representations of decision tables.

G. SUMMARY: FORM AND POSITION, THE TWO FUNDAMENTAL CARRIERS OF INFORMATION

Figure 7.14 is a computer representation of the policy map shown in figure 7.12. It displays a correspondence between condition vectors and action vectors. Where no action vector has been specified, the space reserved for it is left blank. The "insight-seeking" kind of algorithm we discussed in the preceding section for a Veitch-Karnaugh map can be generalized to policy maps of this kind.

As we have seen, visual insight is obtainable only in relatively simple cases. Computer "insight" is possible in all cases. The only limiting factors are time and cost. The range of policy maps for which computer programs are practical, economical tools for the attainment of insight, however, is enormous compared to alternative methods.

The search for "insight" by computer is fundamentally a search for an efficient set of composite rules, using techniques that are computer equivalents of those we described in chapter 4, section C. The search is guided by two fundamental information "carriers": form and position.

In the tables displayed in figure 7.13 and 7.14, the position of a code within a vector tells us which attribute or set it represents. The position of a vector within a table of vectors tells us which rule we are considering. The juxtaposition of a condition vector and an action vector describes the correspondence between them. And the value, or shape, or form of the actual codes within the vectors tell us what value or state of each attribute we are considering in any particular rule.

These two characteristics—form and position—are the only two a digital computer can use to guide its workings. The right form in the wrong position (for example, an invoice number in the slot reserved for purchase-order numbers) can be even more damaging than the wrong form in the right position (for example, an incorrectly calculated invoice amount).

REVIEW

Decision tables can provide the basis for using a digital computer as a "logic machine"—a machine to calculate the implications of logical relationships.

The use of symbols, diagrams, and machines to study logical relationships finds expression in such forms as Boolean algebra, Venn diagrams, Veitch-Karnaugh maps, and truth-value calculators. These are either special-purpose computers, general-purpose computers, or map-and-counter processes such as those devised by Lewis Carroll.

All of these deal with two-valued inputs—usually called (0,1) or true-false—and a two-valued output. Decision tables are similar in structure but generalize the inputs and outputs by allowing more than two permissible values.

The function of logical maps is to facilitate the attainment of insight by means of visual inspection of the maps.

This kind of insight is attainable only in relatively simple cases. For complicated cases with many variables—particularly variables with more than two possible values—logical maps are of limited utility.

For the general extended-entry decision table, the term *policy map* is more descriptive than the term *logical map*.

Policy maps can be represented within a computer in a form that permits the writing of programs to achieve the same kind of "insight," by means of a processing algorithm, as that attainable by visual inspection of logical maps. The fundamental techniques are those described in the earlier chapters of this book, the techniques for consolidating rules and checking a decision table for completeness and consistency.

EXERCISES

1. Assuming the "action-row" coding for logical functions described above (the one in which the logical functions *and, or,* and *implies* are coded as 0001, 0111, 1101), list all sixteen possible logical functions of two variables and give each a descriptive name.

2. All possible combinations of true-false values of two variables, X and Y say, are given by a condition entry whose first row is 0011 and whose second row is 0101. The corresponding values of ~X and ~Y are obtained by applying the function *not* to each condition square. This gives us a first row which is 1100 and a second row which is 1010. Construct a decision table that *ands* the ~X and ~Y rows. Apply the function *not* to the resulting action row. Compare the result with the action row of figure 7.2b. The identity proved by these operations

$$X \vee Y \equiv \sim(\sim X \wedge \sim Y)$$

is known as De Morgan's theorem.

3. In figure 7.10, the expression

$$AB'C'D' + B'C + A'B'C' + AC'D + A'BC'D$$

simplifies to

$$B' + C'D$$

Why is A irrelevant to determining the truth-value of the original expression?

4. Construct a policy map for the decision table of figure 2.4.

5. How many squares would be required in a policy map for the decision table of figure 2.8?

Is Tom tall?	No	No	Yes	Yes
Is Dick tall?	No	Yes	No	Yes
Pout, sulk, call Harry	X			
Call Dick		X		
Call Tom			X	
Heads, Tom; tails, Dick				X

Limited-entry. Two permissible inputs

Two-variable. Up to four possible outputs

7.1a Two-variable limited-entry decision table: Tall girl looking for a partner to invite to Sadie Hawkins dance, with Harry as last resort

Simple propositions	Tom is tall	False	False	True	True
	Dick is tall	False	True	False	True
Compound proposition	Tom is tall and Dick is tall	False	False	False	True

Combinations of true-false possibilities for simple propositions.

Resulting truth or falsity of compound proposition. Only two possible outputs.

7.1b Two-variable truth table for local connective "and": Dependence of assertion (proposition) "Both Tom and Dick are tall" on truth or falsity of individual assertions (propositions) "Tom is tall" and "Dick is tall"

FIGURE 7.1 Comparison of two-variable decision table with two-variable truth table

115

A	0	0	1	1
B	0	1	0	1
A∧B	0	0	0	1

A	0	0	1	1
B	0	1	0	1
A∨B	0	1	1	1

A	0	1
~A	1	0

A	0	0	1	1
B	0	1	0	1
A⊃B	1	1	0	1

	Logical connective	Symbol
	and	∧
(inclusive)	or	∨
	not	~
	implies	⊃

Truth value	Symbol
False	0
True	1

Note: For some purposes, it is more convenient to symbolize "not" by a bar over the symbol being negated (for example, ~A≡Ā≡ not A) or a prime symbol (for example, ~A≡A').

FIGURE 7.2 Truth tables for logical functions: *and, or not, implies*

	1	2	3	4	5	6	7	8
Tall?	0	0	0	0	1	1	1	1
Dark?	0	0	1	1	0	0	1	1
Handsome?	0	1	0	1	0	1	0	1
Preference 1–8	8	7	6	4	5	3	2	1

	1	2	3	4	5	6	7	8
A	0	0	0	0	1	1	1	1
B	0	0	1	1	0	0	1	1
C	0	1	0	1	0	1	0	1
$A \wedge (B \vee {\sim}C)$	0	0	0	0	1	0	1	1

7.3a Three-variable decision table: Girl seeking tall, dark, handsome date but willing to settle for less

7.3b Three-variable truth table: Dependence of truth or falsity of compound proposition $A \wedge (B \vee {\sim}C)$ on truth or falsity of simple propositions A, B, C

FIGURE 7.3 Comparison of three-variable decision table with three-variable truth table

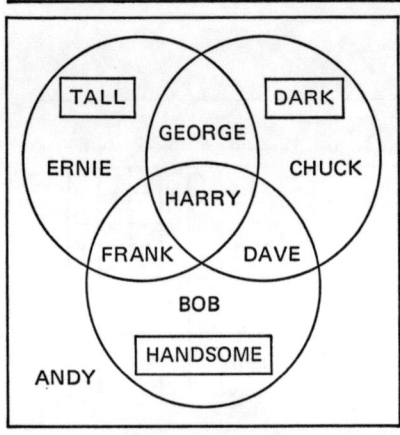 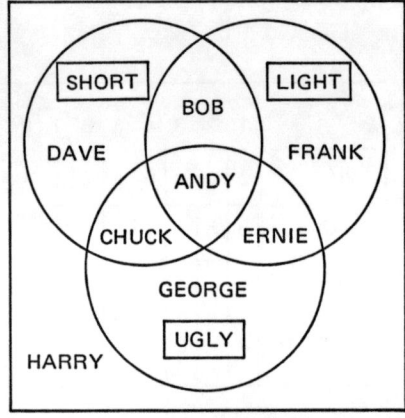

Tall ≡ (Ernie, Frank, George, Harry)
Dark ≡ (Chuck, Dave, George, Harry)
Handsome ≡ (Bob, Dave, Frank, Harry)
Excluded set ≡ (Andy)
Universal set ≡ (Andy, Bob, Chuck, Dave, Ernie, Frank, George, Harry)

Short ≡ (Andy, Bob, Chuck, Dave)
Light ≡ (Andy, Bob, Ernie, Frank)
Ugly ≡ (Andy, Chuck, Ernie, George)
Excluded set ≡ (Harry)
Universal set ≡ (Andy, Bob, Chuck, Dave, Ernie, Frank, George, Harry)

7.4a Venn diagram showing subsets formed by *tall, dark, handsome* sets symbolized by circles

7.4b Venn diagram showing subsets formed by *short, light, ugly* sets symbolized by circles (These sets are the complements of those in 7.4a.)

FIGURE 7.4 Venn diagrams displaying set and subset relationships

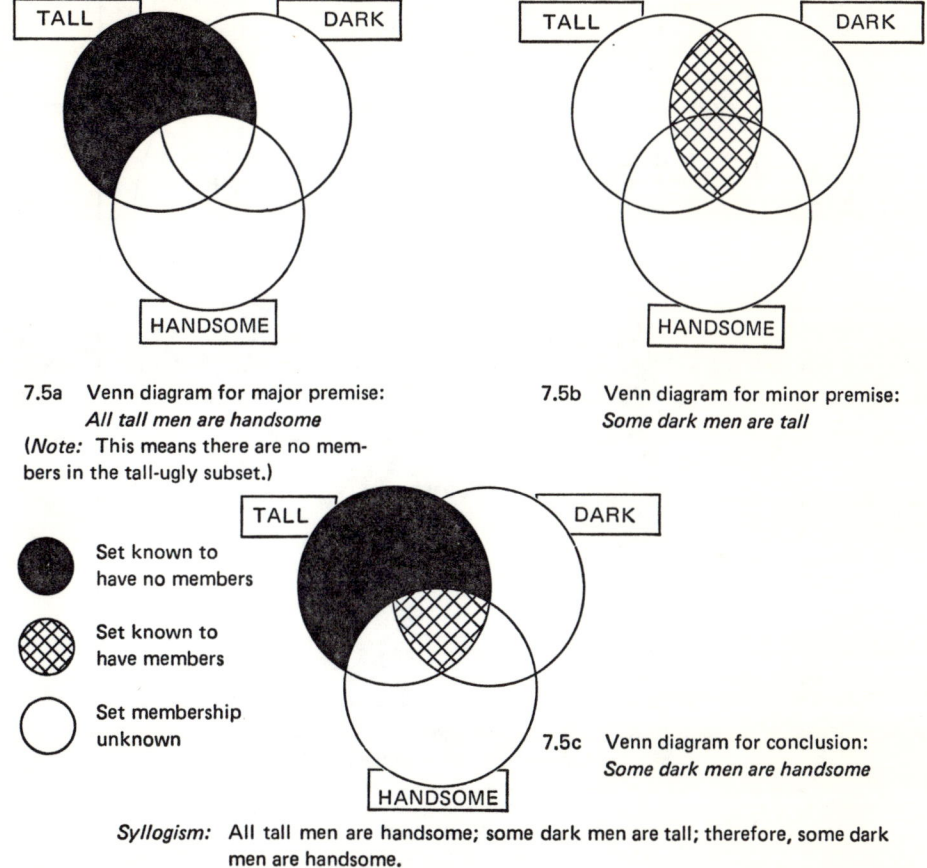

FIGURE 7.5 Use of Venn diagram to prove syllogisms

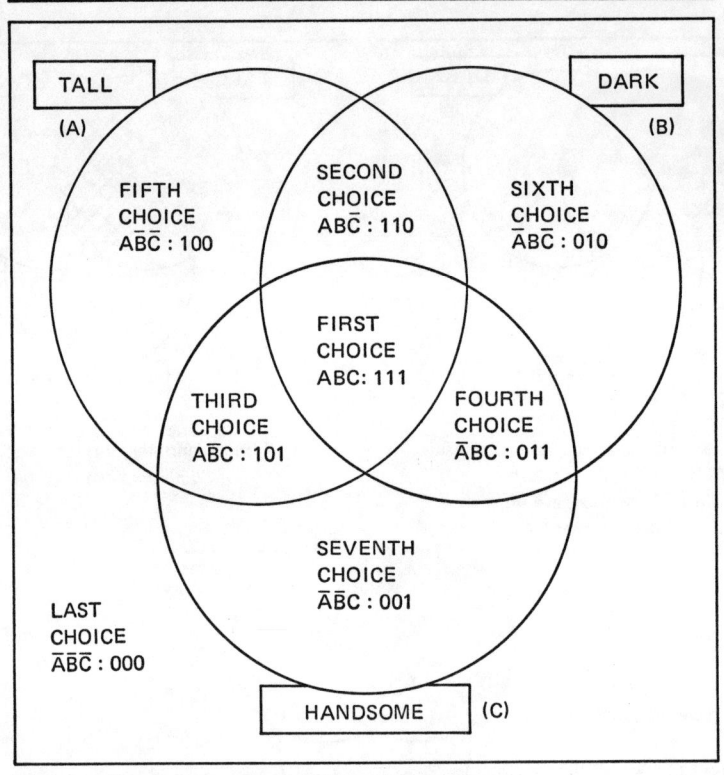

7.6a Venn diagram used to map choices from decision table of figure 7.3a

FIGURE 7.6 Venn diagrams used to map: (a) entry portion of decision table, (b) logical (Boolean) function

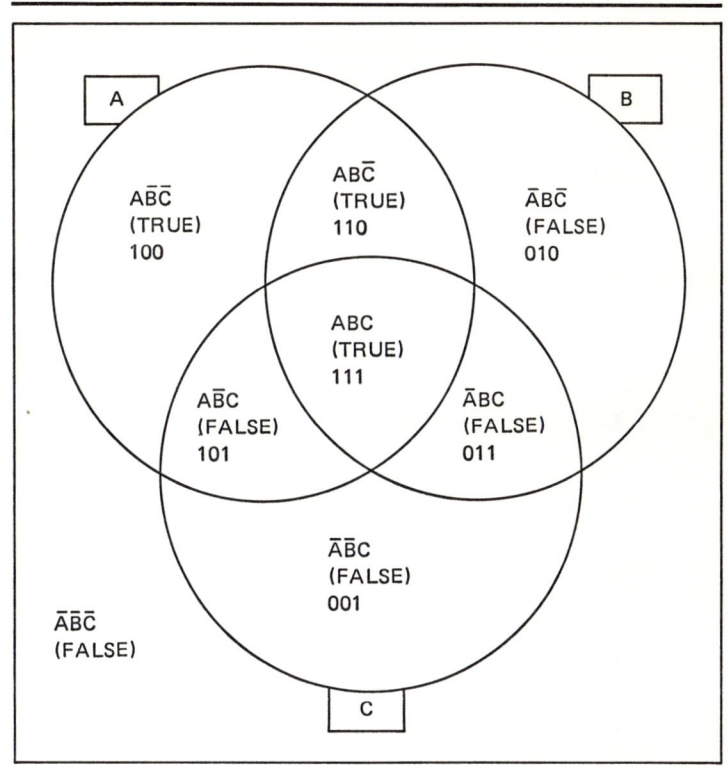

7.6b Venn diagram used to map logical function A ∧ (B ∨ ~C) from truth table of figure 7.3b

FIGURE 7.6 *Continued*

LOGIC	SET THEORY	DENOTED BY:	EXAMPLES
PROPOSITION	SET	1. LETTERS	A, B, C, ...
AND	INTERSECTION MEET (OVERLAP)	2. JUXTAPOSITION 2(A). 'AND' SYMBOL, $\wedge$ 2(B). 'INTERSECTION' SYMBOL: $\cap$	AB, BA, CF $A \wedge B, B \wedge A, C \wedge F$ $A \cap B, B \cap A, C \cap F$
OR	UNION JOIN (COMBINE)	3. PLUS SIGN: + 3(A). 'OR' SYMBOL: $\vee$ 3(B). 'UNION' SYMBOL: $\cup$	A+B, B+A, C+F $A \vee B, B \vee A, C \vee F$ $A \cup B, B \cup A, C \cup F$
NOT	COMPLEMENT NEGATE	4. TILDE: ~ 4(A). OVERBAR: ¯ 4(B). PRIME SYMBOL: '	~A, ~B, ~AB $\bar{A}, \bar{B}, \overline{AB}$ A', B', (AB)'

7.7a Words, symbols, and usage for analogous functions in logic (propositional calculus) and set theory

3-VARIABLE COMPOSITE SUBSETS	EQUIVALENT UNION OF SMALLEST SUBSETS (MINTERMS) (LETTER NOTATION)	UNION OF MINTERMS IN BINARY CODE NOTATION	CONDITION ENTRY NOTATION
A	$ABC \cup AB\bar{C} \cup A\bar{B}C \cup A\bar{B}\bar{C}$	$111 \cup 110 \cup 101 \cup 100$	1XX
$\bar{A}$	$\bar{A}BC \cup \bar{A}B\bar{C} \cup \bar{A}\bar{B}C \cup \bar{A}\bar{B}\bar{C}$	$011 \cup 010 \cup 001 \cup 000$	0XX
B	$ABC \cup AB\bar{C} \cup \bar{A}BC \cup \bar{A}B\bar{C}$	$111 \cup 110 \cup 011 \cup 010$	X1X
$\bar{B}$	$A\bar{B}C \cup A\bar{B}\bar{C} \cup \bar{A}\bar{B}C \cup \bar{A}\bar{B}\bar{C}$	$101 \cup 100 \cup 001 \cup 000$	X0X
C	$ABC \cup A\bar{B}C \cup \bar{A}BC \cup \bar{A}\bar{B}C$	$111 \cup 101 \cup 011 \cup 001$	XX1
$\bar{C}$	$AB\bar{C} \cup A\bar{B}\bar{C} \cup \bar{A}B\bar{C} \cup \bar{A}\bar{B}\bar{C}$	$110 \cup 100 \cup 010 \cup 000$	XX0
$\bar{A}\bar{B}$ or $\bar{B}\bar{A}$	$\bar{A}\bar{B}\bar{C} \cup \bar{A}\bar{B}C$	$000 \cup 001$	00X
$\bar{A}\bar{C}$ or $\bar{C}\bar{A}$	$\bar{A}\bar{B}\bar{C} \cup \bar{A}B\bar{C}$	$000 \cup 010$	0X0
$\bar{A}B$ or $B\bar{A}$	$\bar{A}B\bar{C} \cup \bar{A}BC$	$010 \cup 011$	01X
$\bar{A}C$ or $C\bar{A}$	$\bar{A}\bar{B}C \cup \bar{A}BC$	$001 \cup 011$	0X1
$A\bar{B}$ or $\bar{B}A$	$A\bar{B}\bar{C} \cup A\bar{B}C$	$100 \cup 101$	10X
$A\bar{C}$ or $\bar{C}A$	$A\bar{B}\bar{C} \cup AB\bar{C}$	$100 \cup 110$	1X0
AB or BA	$AB\bar{C} \cup ABC$	$110 \cup 111$	11X
AC or CA	$A\bar{B}C \cup ABC$	$101 \cup 111$	1X1
$\bar{B}\bar{C}$ or $\bar{C}\bar{B}$	$\bar{A}\bar{B}\bar{C} \cup A\bar{B}\bar{C}$	$000 \cup 100$	X00
$\bar{B}C$ or $C\bar{B}$	$\bar{A}\bar{B}C \cup A\bar{B}C$	$001 \cup 101$	X01
$B\bar{C}$ or $\bar{C}B$	$\bar{A}B\bar{C} \cup AB\bar{C}$	$010 \cup 110$	X10
BC or CB	$\bar{A}BC \cup ABC$	$011 \cup 111$	X11

7.7b Composite sets expressed as unions of smallest subsets

FIGURE 7.7 Terminology and notation from logic (propositional calculus), set theory, and decision tables

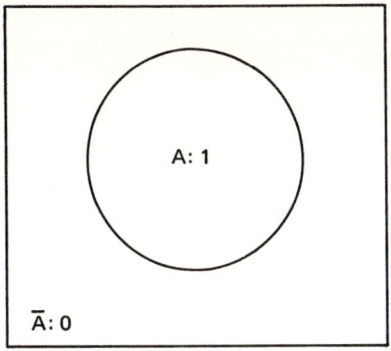

7.8a One variable divides universe into two parts: (0, 1)

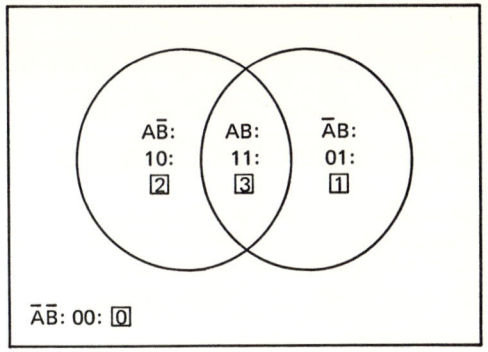

7.8b Two variables divide universe into $2^2 = 4$ parts (0, 1, 2, 3)

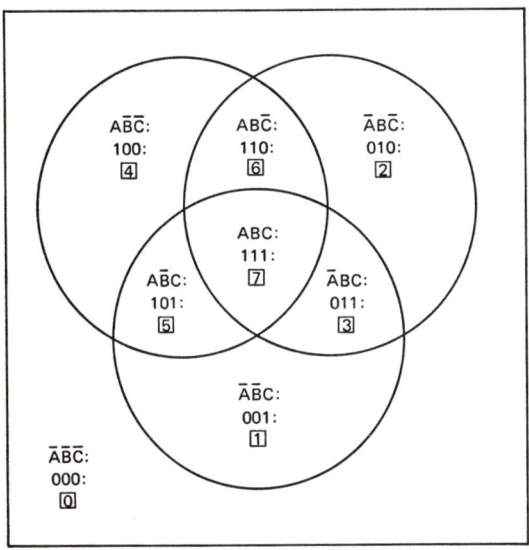

7.8c Three variables divide universe into $2^3 = 8$ parts (0, 1, 2, 3, 4, 5, 6, 7)

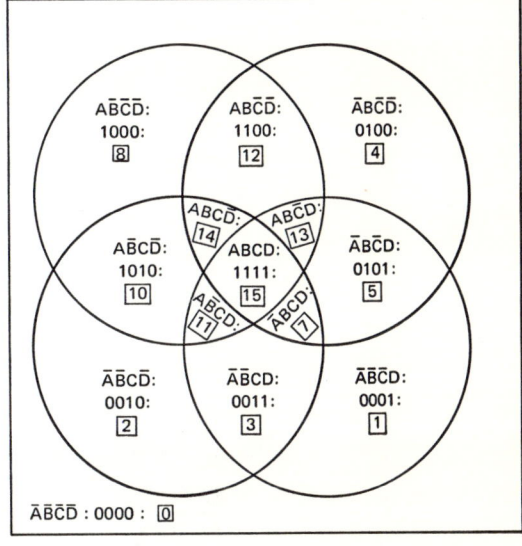

7.8d Four variables divide universe into sixteen parts. Only fourteen can be pictured with intersecting circles. $\bar{A}BC\bar{D}$ and $A\bar{B}\bar{C}D$ missing.

FIGURE 7.8 Structure and limitations of the Venn diagram

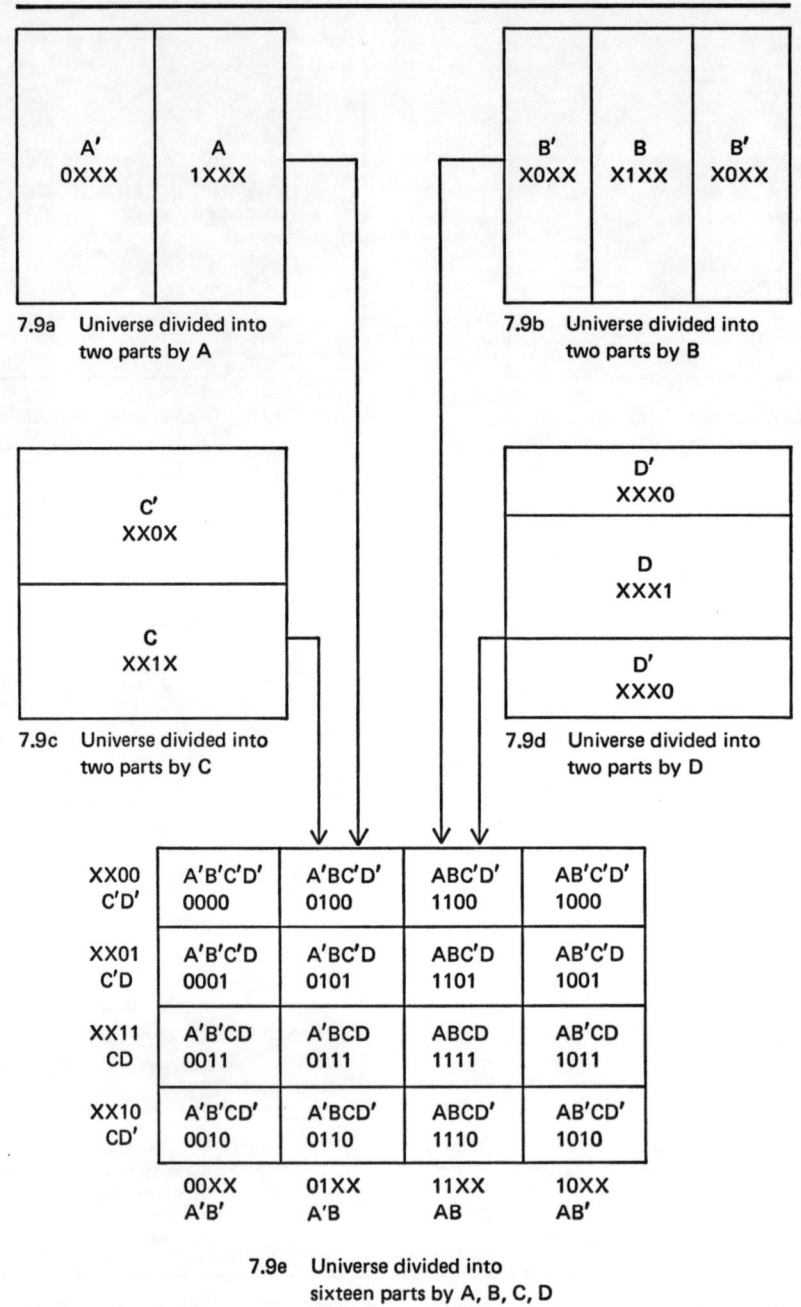

FIGURE 7.9 Four-variable Veitch-Karnaugh map showing subdivision of universe into sixteen minterms

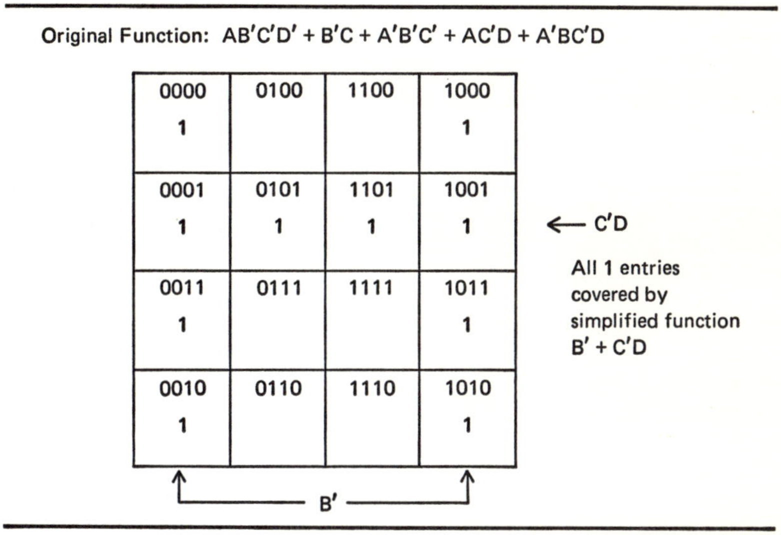

Original Function: AB'C'D' + B'C + A'B'C' + AC'D + A'BC'D

0000	0100	1100	1000
1			1
0001	0101	1101	1001
1	1	1	1
0011	0111	1111	1011
1			1
0010	0110	1110	1010
1			1

← C'D

All 1 entries covered by simplified function B' + C'D

↑——— B' ———↑

FIGURE 7.10 Veitch-Karnaugh map simplifying Boolean expression AB'C'D' + B'C + A'B'C' + A'BC'D

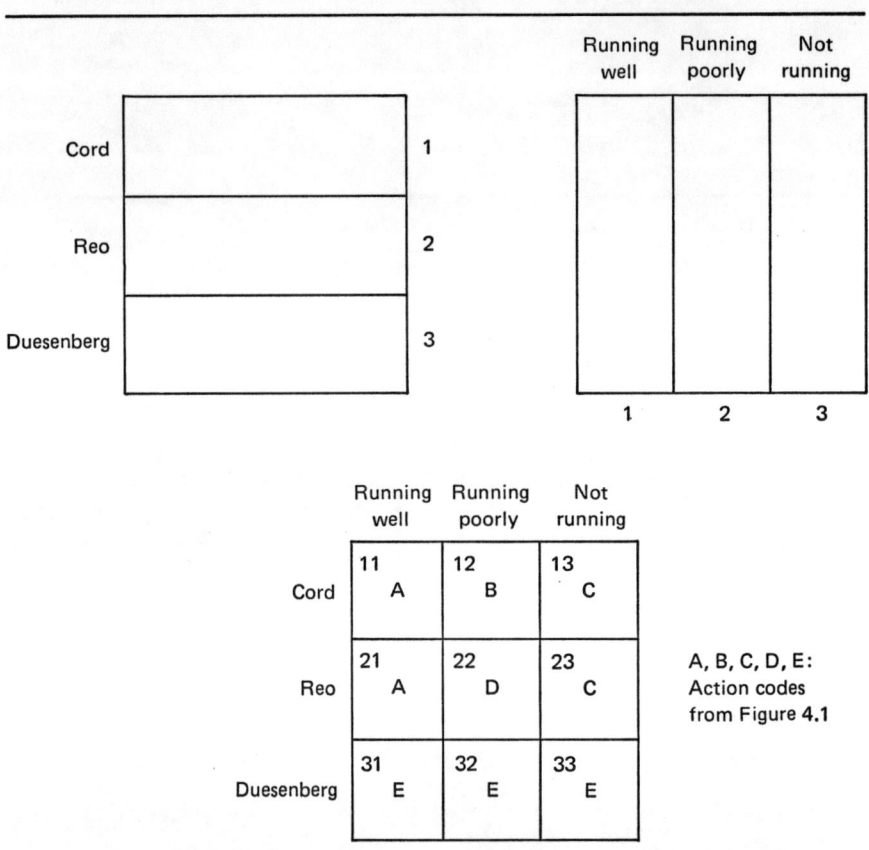

FIGURE 7.11 Integer-valued policy map for vintage-car decision table

Product Class / Invisible Amount →	1. Amt < 10			2. 10 ≤ Amt ≤ 50			3. 50 ≤ Amt ≤ 100			4. 100 < Amount		
Customer Class ↓	1. Engine	2. Pump	3. Fan	1. Engine	2. Pump	3. Fan	1. Engine	2. Pump	3. Fan	1. Engine	2. Pump	3. Fan
1. Retail	111	111	111	111	111	111	111	111	111	111	111	111
2. Government	112	112	112	112	112	112	112	112	112	312	312	312
3. Engine agent		212	212		212	212	623	212	212	723	212	212
4. Pump agent	212	423	212	212	423	212	212	423	212	212	423	212
5. Pump distributor	312	514	312	312	614	312	312		312	312		312
6. Fan distributor	212	212	412	212	212	412	212	212	412	212	212	412

7.12a   Condition policy map

Discount / Consignment Terms →	1. No Consignment				2. Consignment			
	1. COD	2. 30/R	3. 30/S	4. 30-60-90/R	1. COD	2. 30/R	3. 30/S	4. 30-60-90/R
1. 0	1XX	2$\overline{4}$X						
2. 10%		3X$\overline{1}$ 4X$\overline{2}$ 6X$\overline{3}$						
3. 15%		24X 5X$\overline{2}$						
4. 25%		6X3						4X2
5. 30%				512				
6. 33%				522			331	
7. 40%							341	

7.12b   Action policy map

FIGURE 7.12   Policy map for invoicing procedure

```
0 0 0 0 : 1
0 0 0 1 : 1
0 0 1 0 : 1
0 0 1 1 : 1
0 1 0 0 : 0
0 1 0 1 : 1
0 1 1 0 : 0       0 0 0 0 : 1
0 1 1 1 : 0       0 0 0 1 : 1
1 0 0 0 : 1       0 0 1 0 : 1
1 0 0 1 : 1       0 0 1 1 : 1
1 0 1 0 : 1       0 1 0 1 : 1
1 0 1 1 : 1       1 0 0 0 : 1
1 1 0 0 : 0       1 0 0 1 : 1
1 1 0 1 : 1       1 0 1 0 : 1
1 1 1 0 : 0       1 0 1 1 : 1       0 1 0 1 : 1
1 1 1 1 : 0       1 1 0 1 : 1       1 1 0 1 : 1
                  0 : 5 8 6 4       0 : 1 0 2 0
                  1 : 5 2 4 6       1 : 1 2 0 2

Figure 7.13a      Figure 7.13b      Figure 7.13c
```

FIGURE 7.13 Computer representation of Veitch-Karnaugh map of figure 7.10

```
111:111    411:212
112:111    412:423
113:111    413:212
121:111    421:212
122:111    422:423
123:111    423:212
131:111    431:212
132:111    432:423
133:111    433:212
141:111    441:212
142:111    442:423
143:111    443:212
211:112    511:312
212:112    512:514
213:112    513:312
221:112    521:312
222:112    522:614
223:112    523:312
231:112    531:312
232:112    532:
233:112    533:312
241:312    541:312
242:312    542:
243:312    543:312
311:       611:212
312:212    612:212
313:212    613:412
321:       621:212
322:212    622:212
323:212    623:412
331:623    631:212
332:212    632:212
333:212    633:412
341:723    641:212
342:212    642:212
343:212    643:412
```

FIGURE 7.14 Computer representation of invoicing policy map of figure 7.12

8

Division Tables and Flowcharts

A. BRANCHES AND SUBTABLES

In a conventionally drawn flowchart, the lozenges identify *branch points*. The lines leaving the lozenges are *branch lines*. At each branch point, one of the conditions in a condition entry is tested. Each of the branch lines leaving the lozenge corresponds to one or more values of the tested condition; in other words, it corresponds to a rule that is either partially or completely determined.

The effect of the test conducted at a branch point is to divide a table (or subtable) into subtables—that is, simple or composite rules. The division into smaller and smaller subtables continues until subtables are reached that determine a course of action.

The process is illustrated in figure 8.1, which shows the subdivision of the coded vintage-car table of figure 2.3 by a flowchart employing ternary (three-way) branching.

In figure 8.1, the input to the first lozenge in the flowchart is the complete condition entry of figure 2.3. The first lozenge tests the *Make* condition. This test subdivides the original table of nine rules into three subtables of three rules each. The three branch lines associated with these subtables are described by the composite rule codes 1X, 2X, 3X, corresponding respectively to Cord, Reo, Duesenberg. Attached to each line is a diagram of the three simple rules making up its subtable.

The 3X (Duesenberg) line requires no further testing. It leads directly to action set E (described by code 352: variable commission; estimated shopwork; manager O.K. required). Each of the other lines requires a test on *Condition*. This test divides their subtables into three simple rules. The lines corresponding to rules 11 and 21 combine to lead to action set A; the lines corresponding to rules 13 and 23 combine to lead to action set C. Line 12 leads to action set B; line 22 to D.

Note how both incompletely and completely specified rule codes serve as branch labels or "state" indicators. If these labels are appended to the branch lines of flowcharts, they help us to remedy one of the flowchart deficiencies noted in chapter 3, the inability to tell, at any particular point in a complicated flowchart, what tests have been made at preceding branch points.

The potential of these branch labels, either as annotations to flowcharts or as debugging aids when a new program is being tested, is as yet insufficiently exploited. When a program malfunctions, the most useful information a programmer can have in tracking down the source of error is a list of the branch lines his program traversed before the error was detected. If, as part of a checkout procedure, branch labels are developed and stored in standard locations, the "history" of a program that malfunctions can readily be determined.

Most flowcharts do binary (two-way) branching. The principles are the same as for the more general kind of branching illustrated in figure 8.1. Figure 8.2 is a binary-branch flowchart for the same procedure. Like figure 8.1, it displays branch labels and their associated subtables.

B. OPTIMIZING BRANCHING STRUCTURE: FIRST PRINCIPLES

In figure 8.1, the branch points occurred in a "natural" order—first, *Make;* second, *Condition.* In figure 8.2, the sequence of branch points is not "natural." The branch points are, successively, the following questions: Duesenberg? Running well? Not running? Cord? —Why was this sequence chosen?

The reason is suggested by a comparison of figure 8.2 with figure 8.3, a "straightforward" binary-branch flowchart for the same procedure. Along any path in figure 8.3 the first test or tests determine *Make;* the following tests determine *Condition.*

The result is a flowchart with six branch points instead of the four of figure 8.2. The reason for the "unnatural" testing sequence of figure 8.2 should now be clear: it requires fewer branch points, and consequently fewer branching instructions in a computer program based on it. Branch instructions, like other instructions, take up space in a computer's memory or "store." For this simple example, a change in the sequence in which we tested the conditions in our table resulted in saving one-third the space required for branching instructions.

For large, complicated programs the storage savings that can be achieved by a properly designed sequence of branching instructions can be significant. A decision table provides all the information needed to design a flowchart with a minimum number of branch points.

As usual, we are not concerned with details but with a general understanding. How does the order in which we ask questions affect the number of questions we have to ask? Let us compare figures 8.2 and 8.3 to see how the first flowchart achieved its economies.

In figure 8.2, the first question splits the complete table into two subtables: (1) Duesenberg and (2) Cord-Reo. The Duesenberg subtable requires no further subdivision; it leads directly to action set E.

In the Cord-Reo subtable, two values of *Condition (Running Well* and *Not Running)* lead to the same action set for either car. Therefore, if we test for those two values of *Condition* before we test again for *Make,* we can immediately determine what courses of action to follow without testing to differentiate a Cord from a Reo. In fact, we do not differentiate a Cord from a Reo until we have to—that is, for the remaining value of *Condition (Running Poorly),* in which case *Cord* leads to action set B and *Reo* leads to action set D.

In figure 8.3, on the other hand, we differentiate Cord from Reo at the very first branch point. This means that we can no longer take advantage of the fact that, for two out of three values of *Condition,* either Cord or Reo will lead to the same action. We have to repeat identical tests along both branch lines. Therefore, where we had two *Condition* branch points in figure 8.2, we now have four *Condition* branch points in figure 8.3. The difference is clear when we note the number of lines leading to action sets A and C in figure 8.2 and compare it with the number leading to those action sets in figure 8.3.

When a table has composite rules or, more generally, two or more rules with the same action set, the order in which conditions are tested affects the number of tests that have to be made. For complicated cases—those with overlapping composite rules or with several rules that cannot be consolidated but have the same action set—a computer program should determine the testing sequence. In most situations, however, a simple principle will achieve significant economies in the space required for branching instructions:

> *Avoid or postpone tests which will divide the subtables that correspond to composite rules in the decision table.*

This principle was used in selecting the test sequence of figure 8.2.

C. SAVING SPACE OR TIME OR BOTH

In many cases, the branch sequence with the fewest number of branch points will achieve not only storage economy but also economy in execution time. This is because, in general, the fewer instructions you store, the fewer you execute.

This is not always the case, however. If some of the rules in a decision table are invoked more frequently than others, then to assure minimum execution time, those rules should be determined as quickly as possible—

that is, the numbers of branch points on the branch path leading to the frequently occurring rules should be as small as possible.

In figure 8.2, the numbers of tests required to determine which of 5 courses of action to pursue are as follows:

> E: 1
> A: 2
> C: 3
> B: 4
> D: 4

In a program based on this flowchart, the number of branching instructions that the computer will execute during the life of this application will depend on the percent of sales in each of the nine categories in our table.

If the agency sells nothing but Duesenbergs, for example, the program will never execute more than one branching instruction. If, on the other hand, the agency sells nothing but poorly running Cords and Reos, the program will always execute four branching instructions.

In figure 8.4, we show the effects on branch-instruction execution of two different sales patterns: situations I and II.

In situation I, all sales possibilities are equally likely. This means that the relative frequencies with which the nine simple rules of our table will occur is a little over 11 percent. These frequencies are shown as decimal fractions in column I of figure 8.4a.

In situation II, 80 percent of the sales are of the two *Make-Condition* categories that have the longest branch paths in figure 8.2, the *Cord-Running Poorly* and *Reo-Running Poorly* cases. The detailed list of decimal fractions for these and the other cases is shown in column II of figure 8.4a.

We determine the average number of branch executions for either situation by multiplying the relative frequency of a rule by the number of branch instructions in its branch path and adding up the results. In situation I, for example, rule A occurs 22 percent of the time, since it consists of two simple rules each of which occurs 11 percent of the time. In situation II, rule A occurs only 7 percent of the time. For both cases, we multiply the corresponding decimal fractions by 2, since there are two branch points in the path leading to action set A.

The remaining rules are treated the same way. The detailed calculations are shown in figure 8.4b and 8.4c. The situation I average is 2.32; the situation II average is 3.61. When the sales percentages are those of situation II, a program based on figure 8.2 will, on the average, spend over 50 percent more time executing branching instructions than it will when the sale percentages are those of situation I.

For frequently used programs, a branching structure that trades storage space for savings in execution time might be advisable. The information provided by the decision table itself is not sufficient to permit design of a minimum-execution-time branching structure. It has to be supplemented with frequency information similar to that given in figure 8.4.

Figure 8.5 is a flowchart that reduces the average number of branches executed for situation II from 3.61 to 2.60 at the expense of adding two more branch points. It does this by concentrating on the rules that occur most frequently and performing the tests that will isolate those rules as quickly as possible. As the caption to the figure implies, the flowchart of figure 8.5 is still not minimum time. The determination of a minimum-execution-time flowchart is left as an exercise for the reader.

An important practical point: the use of rule-frequency information in constructing a minimum-time program is advisable only if it is known that the frequencies will not change from one reporting period to the next. If they change, the situation for which the program was designed no longer applies; in this case, execution times could well be worse than those for a computer program than minimized storage space without regard for execution time. Unless we are justified in assuming that the relative frequencies of the rules will remain the same for a long time, we are better off disregarding them and designing a program that requires as little storage as possible.

Cases where it is advisable to consider rule frequencies are not very common. Such frequencies are usually either not known or not constant. Changing a program to reflect changes in rule frequencies requires reassembling or, in high-level languages, recompiling the program. If this is done even once, it is apt to wipe out any time savings achieved by the faster running times of the optimized code. Frequent recompiling also has other drawbacks: scheduling difficulties, documentation difficulties, introduction of error possibilities, expense, and so on. In general, it can be said that minimum-branch-time programs should only be attempted for applications that are frequently run and characterized by statistical constancy. Such applications are far from the general case.

D. INTERLEAVING TESTS AND ACTIONS

The storage economies to be achieved by following the advice given in section B above are not the only ones possible. Basically, the advice given in that section was to avoid or postpone any tests that introduce branch paths along which the same conditions have to be tested in both paths.

Sometimes identical tests or test patterns occur along separate paths that have not been caused by splitting a composite rule table. Figure 3.3 is an example. The paths leading to rule sets (II, III, IV), (VI, VII, VIII), and (XII, XIII, XIV), all end in two identical tests leading to the same sequence of actions: B, C, D. Clearly, four of the six branch points are unnecessary. All three of the branch lines that precede these two tests could be joined into a single line terminating in a common set of two branch points. In other words, instead of having separate lozenges for the (VI, VII, VIII) and

(XII, XIII, XIV) paths, those lozenges could be removed and their respective branch lines directed to the first of the last two lozenges in the (II, III, IV) path.

This, of course, complicates the task of associating a decision-table rule with a clearly defined path in a flowchart. It also complicates the branch-labeling procedure that was recommended earlier in this chapter. In fact, it brings up a fundamental question that we have so far carefully avoided: the question of how decision tables compare with flowcharts whose branch lines recombine after they have been separated at preceding branch points.

In most flowcharts, this is the common case. The two or more branch lines that leave a branch point normally take individual actions, possibly even make individual tests, and then come together for further testing and action. This is not the situation considered in most books on decision tables, including the earlier chapters of this one.

Most of the examples in decision-table books show long sequences of tests followed by long sequences of actions. A "standard-form" decision table (the kind we have been considering) shows to best advantage when it describes procedures of this kind. Unfortunately, most procedures are *not* of this kind. In most procedures, test sequences and action sequences are relatively short: one or two tests, followed by one or two actions, followed by one or two tests. If we follow the decision-table principles we have so far discussed, we have no choice but to describe such procedures as a large collection of small decision tables with one or two condition rows. If we do this, the table's advantage over the flowchart is not as striking as we have so far made it out to be—in fact, the flowchart may well have the advantage in many cases.

When branch lines recombine, the flowchart achieves economies of description that the decision table in "standard" form cannot achieve. The point can be illustrated by a flowchart in which there is a great deal of branching, taking action, and recombining.

Figure 8.6 is an example of this kind of flowchart. It describes a simple, extremely artificial procedure for determining the value of any number between 0 and 1023 by asking exactly ten yes-no questions, in each case taking action if the answer is yes. Preceding the first question, the variable Y, whose value at the end of the calculation will be the number we are seeking, is initialized to 0. We use the letter X to represent the number about which we are asking questions. The first question is, "Is the quotient of X divided by 512 (disregarding the remainder) odd?" If the answer is yes, the value of Y is increased by 512. If the answer is no, no action is taken; we branch to the next test. The remaining nine tests and actions are similar; they differ only in the divisors used to calculate the quotient about which the question is asked and the amount added to Y if the answer is yes. The remaining nine divisors are 256, 128, 64, 32, 16, 8, 4, 2, 1.

The flowchart is simple and compact. What would the corresponding decision table look like?

It would have 1024 simple rules, the first leading to the "action" (result) 0, the last leading to the result 1023. There are 1024 distinct actions, so no composite rules are possible. The trivially simple flowchart corresponds to a decision table of unmanageable size. If we put 50 rules on a page, it would take 21 pages to replace figure 8.6 with a decision table.

Although the example is artificial, the point it makes is an extremely practical one. Few real procedures separate easily into a long sequence of conditions followed by a long sequence of actions. They can all be *described* this way, of course—just as we could describe each of our 1024 numbers by its own individual rule in a decision table—but such descriptions are almost sure to be unsatisfactory for a variety of reasons: cost, size, incomprehensibility, time required for preparation and maintenance, and so on.

Have we been chasing a wild goose? Has all our study been for naught? Have we been studying the merits of something which has no realistic application?

No. Decision tables can be written even more compactly than flowcharts, including "worst-case" flowcharts with many recombining branches. What is required is a modification of the decision-table conventions we have so far studied.

E. INTRATABLE BRANCHING; ITERATIVE AND RECURSIVE TABLES; ACTION-SEQUENCE NUMBERING

Figure 8.7 is a table similar in form to the standard-form tables discussed in the earlier chapters of this book, except for additional rows of information at the top and bottom of the table. It describes the ten-questions procedure of the preceding section.

The added top row consists of *Subtable* identifiers. The first subtable is identified as "Entry," the remaining ten by the letters A-J.

The Entry subtable consists of a single column that describes the actions required to initialize the table—that is, the actions to be performed before any tests are made. In this case, only one initializing action is required: setting the initial value of Y to zero. Y is the variable (the symbolic name of a storage location in the computer) that will have the output value we seek when the procedure has been carried out.

Subtables A-J correspond to the ten branch points of figure 8.6. Each is, in effect, a decision table with one condition row and two action rows. The first action row changes the value of Y; the second identifies the next subtable to be activated. The first action is taken only for "Yes" or "1" condition entries; the second is *always* taken.

The last action row of the table is a "Branch to" row. For the "Entry" subtable and subtables A-I, it indicates which subtable to activate next. For the "J" subtable it indicates that the procedure is at an end.

By use of the subtable convention, the original 1024 rules required to describe the ten-questions procedure are reduced to 20. Each (0,1) subtable entry corresponds to a branch point in figure 8.6. The space required for a diagram of the decision table is, if anything, a little less than the space required for the flowchart, yet we have retained all the comparative advantages of the decision table that we have stressed in earlier chapters.

The ten-questions procedure is typical of the mathematical procedures commonly called algorithms. It has recurring regularities that permit us to define it iteratively—that is, by repeated execution of the same relatively small number of instructions for different values of its input and output variables. This, however, is not ordinarily characteristic of clerical procedures. That is why we have chosen the noniterative form of flowchart and decision table given in figures 8.6 and 8.7. In the procedures of primary interest to us, the actions in the action stub would not have the identity of form that characterizes the actions in the action stub of figure 8.7.

For iterative procedures, much more economical descriptions are possible in both flowchart and decision-table forms. Figure 8.8 shows an iterative decision table for the ten-questions procedure. The entry subtable initializes two values: Y (the output variable) and N (the divisor). There is only one other subtable: A. A consists of two condition rows and three action rows. The first of the action rows increases the value of Y by N if $|X \div N|$ (the quotient of X divided by N) is odd. The second redefines N to have half its former value. The third branches either to the subtable itself or to the end of the procedure. The procedure ends when the value of N is less than 1.

Figure 8.8 introduces a new and important idea. So far, action squares have been used only to indicate whether the associated action should or should not be taken (limited-entry tables) or which one of a particular set of actions in the corresponding row of the action stub is to be taken (extended-entry tables).

In figure 8.8, a limited-entry table, we introduce a new idea. The entry in an action square indicates not only that the action is to be taken but also *when* it is to be taken with respect to other actions.

The reason for indicating action sequence can best be understood by examining rule 3 of figure 8.8. In this rule, the *Quotient* test indicates that Y should have its value increased by N. But in columns 2 and 3, we divide N by 2 to give it the value needed for the next iteration. Clearly, we have to increase Y by N *before* we divide it rather than after. Therefore in figure 8.8, we put a 1 in the $Y \leftarrow Y + N$ row and a 2 in the $N \leftarrow N \div 2$ row to indicate the order in which these two actions are to be performed.

It is important to realize that everything we have said about checking decision tables for completeness, consistency, and redundancy applies to these new types of decision tables in exactly the same way as it applies to the "standard" forms discussed earlier. The new forms provide more economical, more comprehensible, more efficient descriptions, but they have exactly the same information content. Consolidating, expanding, analyzing can be done just as readily for them as for the larger, explicit tables that can be derived from them.

F. SUMMARY: TABULAR TECHNIQUES

We have, in this chapter, discussed only a few of the many possible ways in which the decision-table idea can be extended to achieve greater efficiency and usefulness for the purpose of systems analysis and programming. This is clearly all we can do in an introductory book.

But the reader should remember that this technique, like all other computer techniques, is in its infancy. The whole question of tabular techniques is one which requires research and development. Its potential is great. Significant economies and performance improvements are attainable now. But truly major changes lie in the future. These changes will affect the day-to-day activities of the systems analyst and programmer. They will come about only through persistence, imagination, and hard work applied to the practical problems faced by these two professions in their day-to-day activities.

REVIEW

If when we interpret a flowchart we keep the structure of an equivalent decision table in mind, we can describe the branch points in the chart as input-output operations. The input is a composite rule describing a complete table or subtable. The output is a division of the input subtable into two or more subtables, each of which is associated with (and described by) one of the branch lines leaving the branch points.

The subtable associated with each line can be described compactly by the digitized form of the composite rule to which it corresponds. These rule descriptions can serve as branch labels that indicate, for any line, what its branching history has been. As state indicators, they can be extremely useful debugging aids.

When decision tables have composite rules, the order in which conditions are tested can affect the number of branch points in the corresponding flowchart and, as a result, the number of branching instructions in the resulting program. A minimum-storage branching sequence can be determined from the information stored in the decision table.

When the rules in a table do not occur with the same relative frequency, then the number of instructions executed by a program will depend on the branching structure of the program. If relative rule frequencies are known, they can be supplied to a decision-table translator for calculation of a minimum-execution-time branching structure.

The decision tables we have so far considered all have one characteristic in common: they continue to branch until a course of action is determined. This is *not* characteristic of most practical procedures. The branch lines of most procedures direct the flow to an action sequence. After this is completed, they rejoin the other branch lines. This can give the flowchart a great advantage over the standard-form decision table in conciseness and comprehensibility.

For procedures characterized by branch lines that recombine after branching, the relative advantages of the decision table over the flowchart can be retained by a modification of the standard-form decision table. In this modification, the basic table structure is unchanged, but the columns are divided into subtables identified by labels above the first rule number in the subtable. The rows associated with the columns are then identifiable by the presence of entries in the condition or action squares. One of the subtables is a special one that will occur in most procedures: the "Entry" or "Initialization" subtable. This subtable tests no conditions; it specifies the actions that are to be performed before any condition is tested.

The addition of these conventions can be extended in a straightforward way to allow for the description of iterative procedures.

EXERCISES

1. Add branch labels to the flowchart of figure 3.3. What do the decimal numbers under or to the left of each branch line signify?

2. Draw figure 3.3 so that it has as few branch points as possible.

3. Draw figure 8.5 so that it will execute the fewest possible branching instructions for the rule frequencies described as situation II in figure 8.4.

4. Of the flowcharts in figures 8.1, 8.2, 8.3, and 8.5, which gives the best performance for situation I? Which the best performance for situation II?

5. Redraw the "Payroll" decision table of exercise 2.7 in a more concise form, using the ideas described in section 8.E.

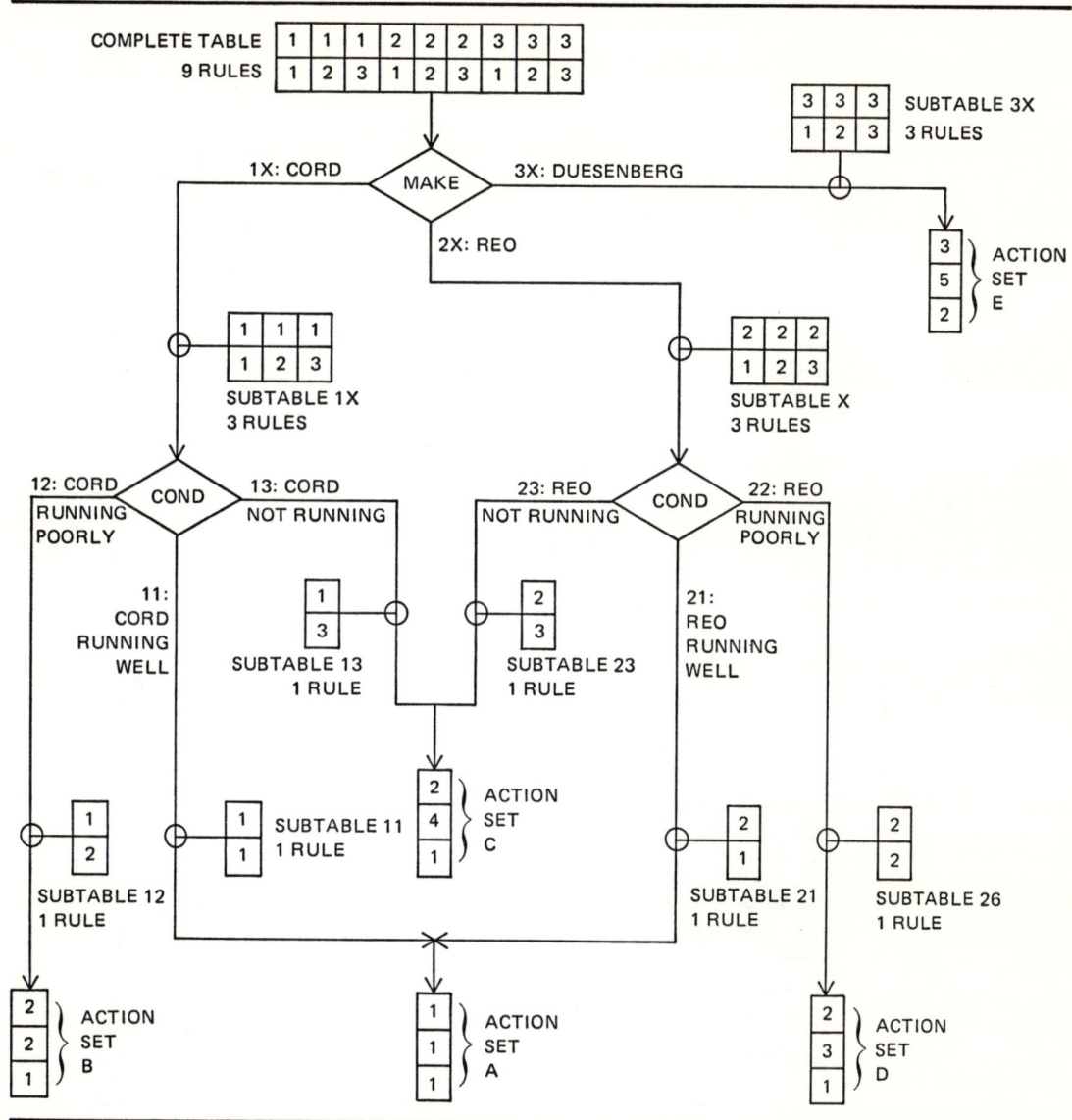

FIGURE 8.1 Flowcharts, subtables, branch labels I: Ternary branching (table of figure 2.3)

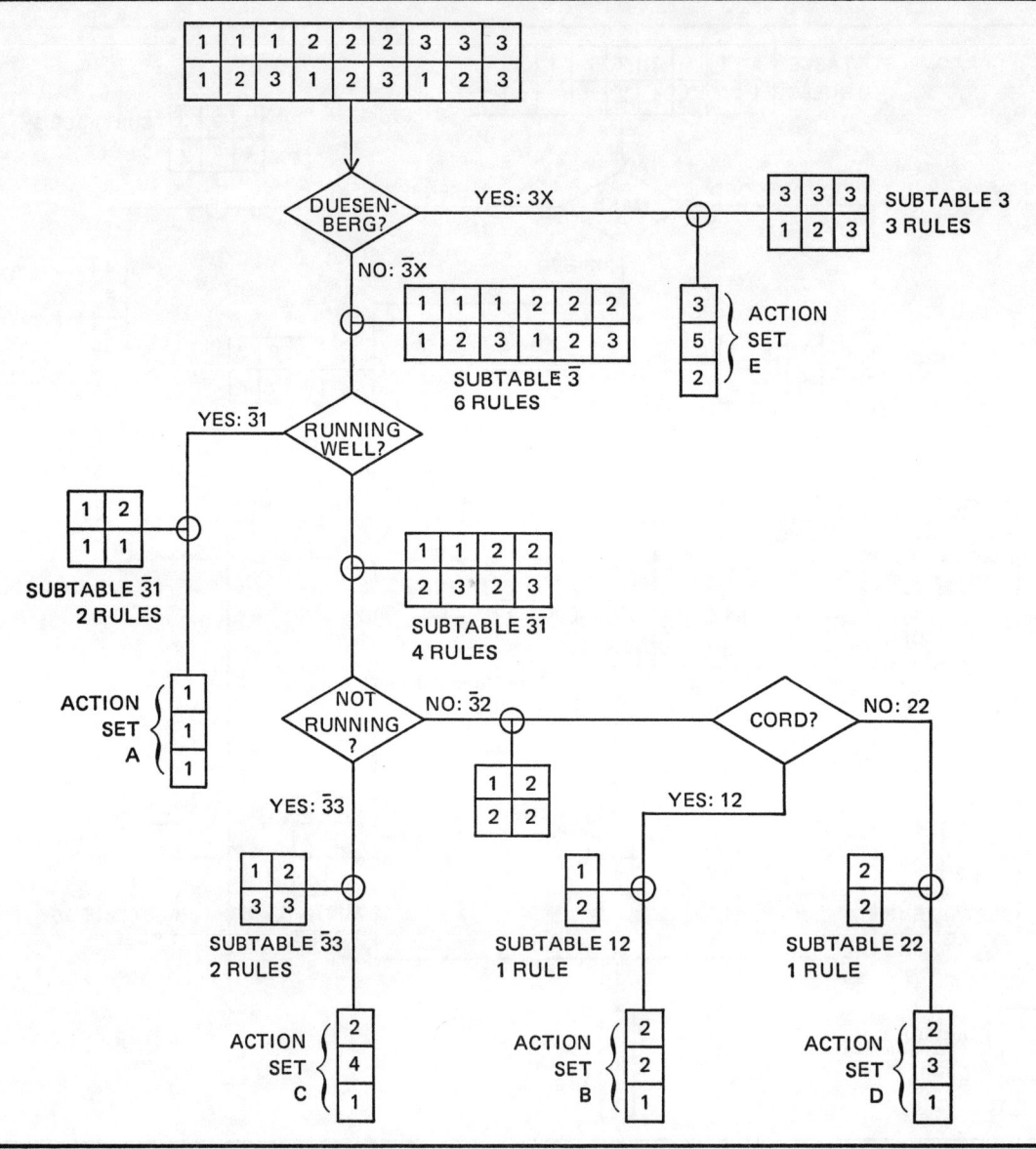

FIGURE 8.2 Flowcharts, subtables, branch labels II: Binary branching (table of figure 2.3)

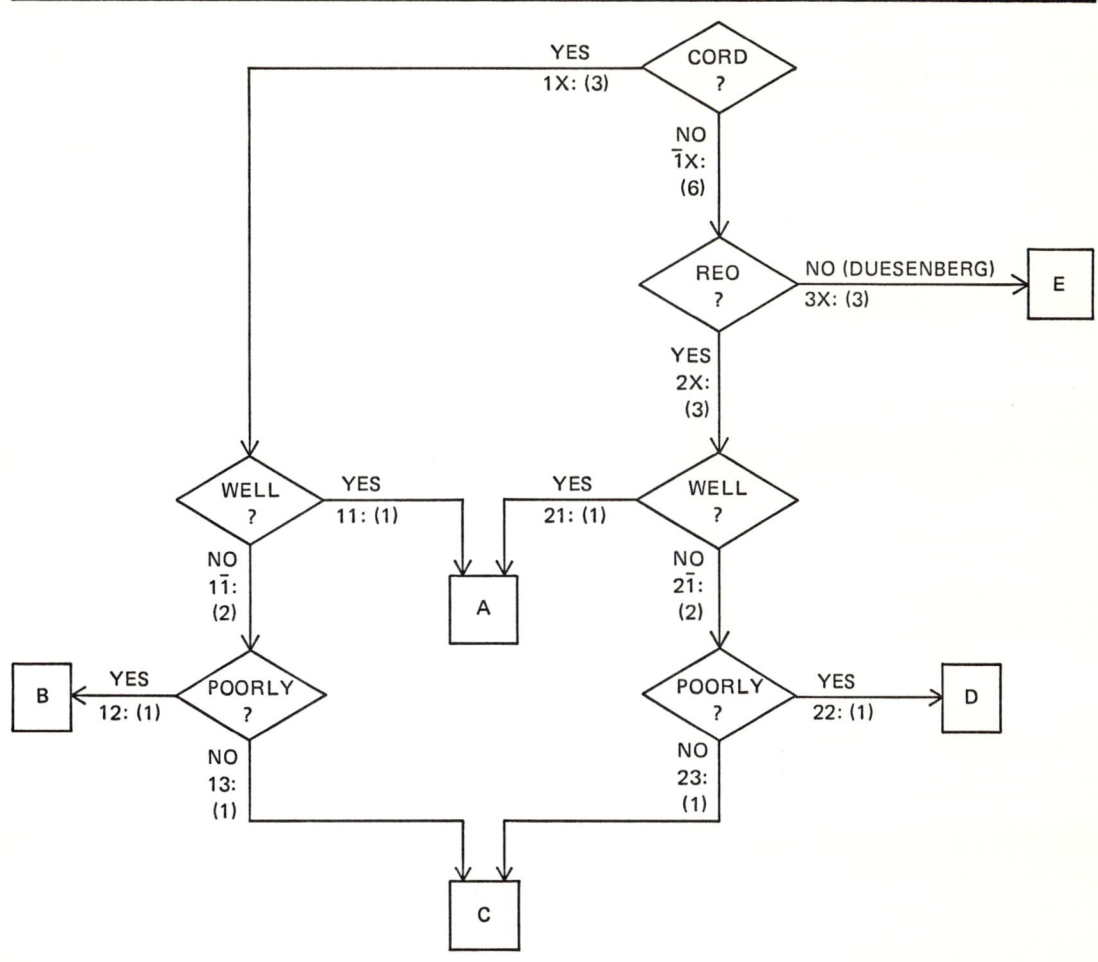

FIGURE 8.3 "Straightforward" version of flowchart of figure 8.2

PERCENT OF SALES: TWO SITUATIONS

	I	II	
Cord, running well	.11	.035	
Cord, running poorly	.11	.40	
Cord, not running	.11	.035	
Reo, running well	.11	.035	
Reo, running poorly	.11	.40	8.4a
Reo, not running	.11	.035	
Duesenberg, running well	.12	.02	
Duesenberg, running poorly	.11	.02	
Duesenberg, not running	.11	.02	

AVERAGE NUMBER OF BRANCH EXECUTIONS: SITUATION I

Duesenberg	$1 \times .34 = .34$	
(Cord, Reo) running well	$2 \times .22 = .44$	
(Cord, Reo) not running	$3 \times .22 = .66$	8.4b
Cord, running poorly	$4 \times .11 = .44$	
Reo, running poorly	$4 \times .11 = .44$	

$\boxed{2.32}$ Average no. of executions

AVERAGE NUMBER OF BRANCH EXECUTIONS: SITUATION II

Duesenberg	$1 \times .06 = .06$	
(Cord, Reo) running well	$2 \times .07 = .14$	
(Cord, Reo) not running	$3 \times .07 = .21$	8.4c
Cord, running poorly	$4 \times .40 = 1.60$	
Reo, running poorly	$4 \times .40 = 1.60$	

$\boxed{3.61}$ Average no. of executions

FIGURE 8.4 Average number of branch-instruction executions: Flowchart of figure 8.2

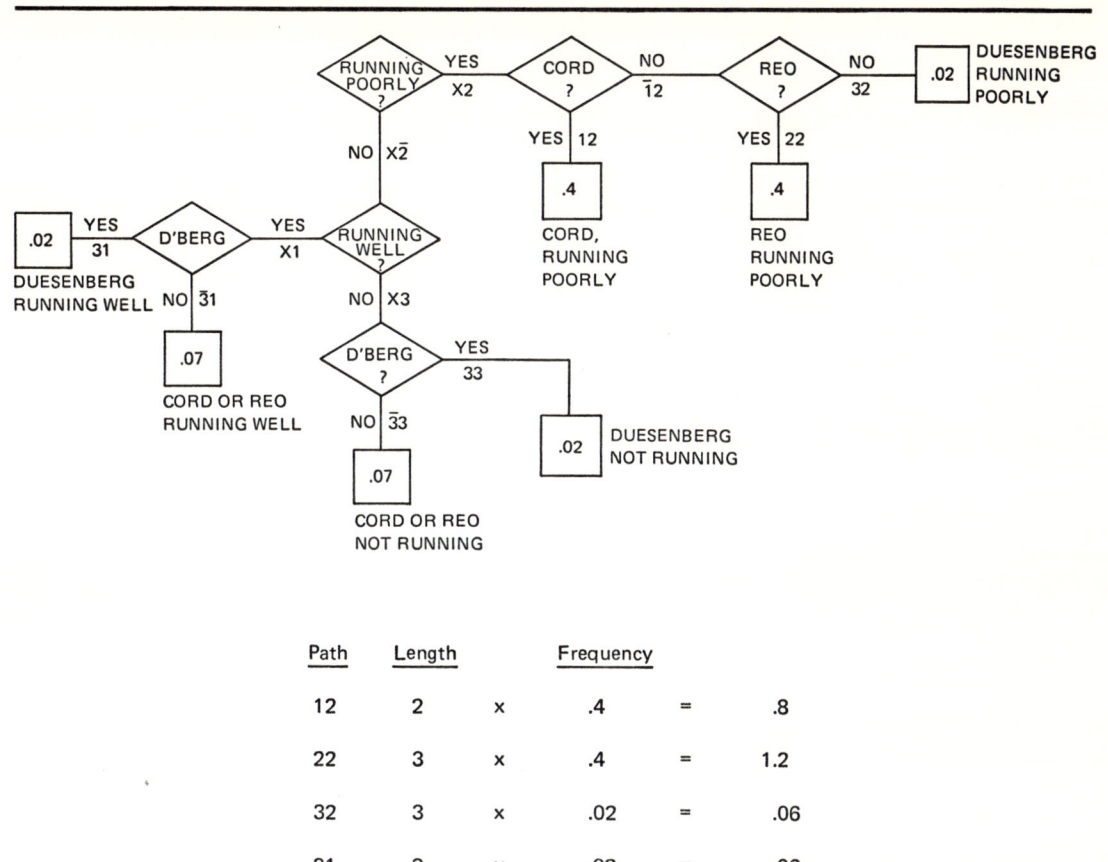

FIGURE 8.5 Improved flowchart for situation II: Is it optimal?

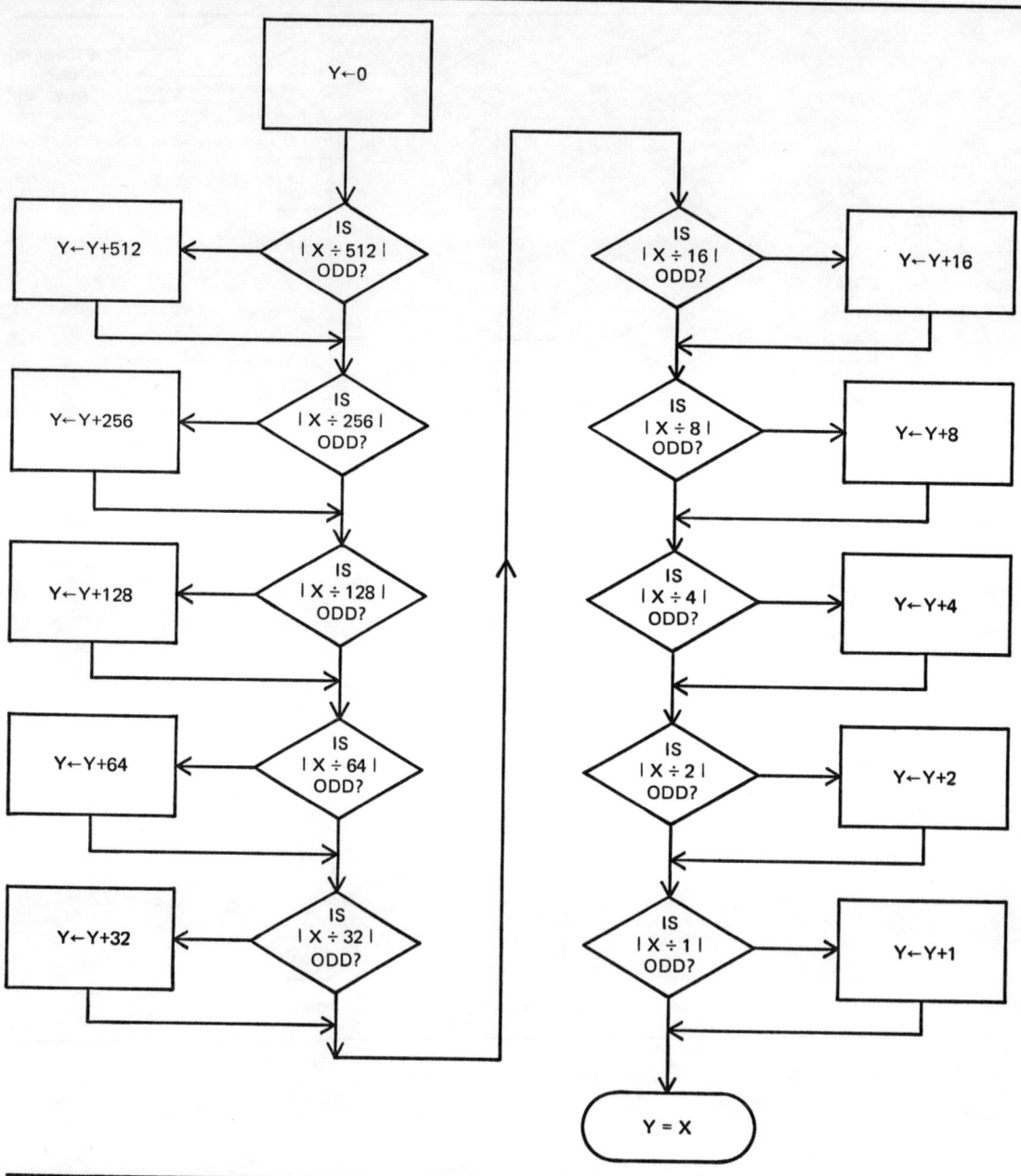

FIGURE 8.6 Flowchart to determine the value of unknown X ($0 \leq X \leq 1023$) by asking ten yes-no questions

Subtable Identifiers	Entry	A		B		C		D		E		F		G		H		I		J	
Rules	1	2	3	4	5	6	7	8	9	10	11	12	13	14	15	16	17	18	19	20	21
Is | X ÷ 512 | odd?		0	1																		
Is | X ÷ 256 | odd?				0	1																
Is | X ÷ 128 | odd?						0	1														
Is | X ÷ 64 | odd?								0	1												
Is | X ÷ 32 | odd?										0	1										
Is | X ÷ 16 | odd?												0	1								
Is | X ÷ 8 | odd?														0	1						
Is | X ÷ 4 | odd?																0	1				
Is | X ÷ 2 | odd?																		0	1		
Is | X ÷ 1 | odd?																				0	1
Y ← 0		X																			
Y ← Y + 512			X																		
Y ← Y + 256					X																
Y ← Y + 128							X														
Y ← Y + 64									X												
Y ← Y + 32											X										
Y ← Y + 16													X								
Y ← Y + 8															X						
Y ← Y + 4																	X				
Y ← Y + 2																			X		
Y ← Y + 1																					X
Branch to		A	B	B	C	C	D	D	E	E	F	F	G	G	H	H	I	I	J	J	End	End

FIGURE 8.7 Intratable branching: The "ten-question" procedure described by an extended form of decision table

Subtable Identifiers →	Entry	A		
Rules →	1	2	3	4
Is N < 1		0	0	1
Is | X ÷ N | odd?		0	1	X
Y ← 0	X			
N ← 512	X			
Y ← Y + N			1	
N ← N ÷ 2		1	2	
Branch to:	A	A	A	End

FIGURE 8.8 Intratable branching, action-sequence numbering: Iterative decision-table conventions

9

Decision–Table Translators

A. FROM DECISION TABLE TO COMPUTER PROGRAM

Having prepared a decision table, we ordinarily want to produce a computer program for the procedure it describes. How can this be done?

Here are four possible ways:

1. Give the decision table to a programmer and have him write a program.
2. Prepare a flowchart from the decision table, give the flowchart to a programmer, and have him write a program.
3. Convert the table into a set of statements in a computer programming language. Feed a suitable form of these statements to the corresponding language translator and have it produce a program.
4. Provide a description of the decision table to a translator program that can translate the description directly into a program.

With the obvious exception of the first, a computer can be used to mechanize all the processes listed above. In practice, however, most existing decision-table translators perform the function described in 3 above—that is, they translate a decision table into a COBOL, FORTRAN, or other high-level language program. This type of decision-table translator is called a *preprocessor*. (The processor, in this case, is the COBOL, FORTRAN, etc. translator itself.) The assistance most commonly provided to an assembly-language programmer—a programmer working with symbolic representations of the instructions characteristic of a particular computer—is one or more decision-table *macros*. These permit him to enter decision-table descriptions into his program at appropriate points. The assembler (translator program) then converts these into the appropriate sequence of machine codes. Unlike the preprocessors, macros are more an *aid* to programming than an alternative form of programming.

A book that attempts to describe the current status of software offerings is almost sure to be out of date by the time it appears in print; the successful programs will by then have proliferated beyond recognition and the unsuccessful will have disappeared. The reader interested in learning the current details of what decision-table translators are commercially available can best get such information from—

1. Computer industry trade publications.
2. The publications of professional computer groups.
3. Representatives of software and hardware vendors.

Our concern, in any event, is not with the details of specific translators but with an understanding of what the translation process accomplishes. In this chapter we attempt to attain such an understanding by—

1. Discussing how programming languages can be written in decision-table format.
2. Illustrating and discussing the workings of an "interactive" translator program.

B. WRITING PROGRAMMING LANGUAGES IN DECISION-TABLE FORMAT

The statements in programming languages perform a variety of functions: describing the structure of files, manipulating the use of storage within a computer, controlling the copying of information from and to peripheral devices, performing calculations, performing tests to determine which calculations to perform next, and so on.

The full range of statements can, for our purposes, be divided into two groups: (1) procedural statements and (2) descriptive (nonprocedural) statements. Decision-table translators are concerned primarily with statements in the first group. The only descriptive statements they ordinarily employ are those required to identify the table itself.

This means that, when a decision-table preprocessor is used in conjunction with a particular language, the nonprocedural portions of the program (for example, the Identification, Environment, and Data divisions of a COBOL program; the FORMAT, DIMENSION, TYPE, COMMON, EQUIVALENCE, etc. statements of a FORTRAN program) are written just as they would be for a program that did not use decision tables. The decision table replaces only the procedural part of the program.

For any programming language, the procedure statements can be divided into two types: (1) branching statements and (2) action statements.

The branching statements can be either conditional or unconditional.

The conditional branches are commonly of the IF . . . THEN form (though many variations are possible). The action statements cover all the actions a computer can take (perform a sequence of calculations, identify the result with a name, and store it for later use; copy information from some external source, give it a name, and store it; copy named information to some external device; initiate a repetitive set of actions; activate a subprogram; branch unconditionally; etc.).

When a preprocessor is used, the conditional branch statements (with the bracketing IF . . . THEN delimiters removed) are written in the condition stub. The action statements are written in the action stub. All possible outcomes of the condition tests are written in the condition entry. An indication of which set of actions to take is written in the action entry.

In effect, what the preprocessor permits is the separation of conditional branch statements from action statements in the way we have seen to be characteristic of the decision-table approach to procedure description. The statements themselves (except for the implied IF . . . THEN or its equivalent) are those of the language in which the program is written (COBOL, FORTRAN, PL/I, BASIC, etc.).

The task of the preprocessor is thus to take the statements that have been separated, supply the appropriate conditional delimiters and interleave branching statements and action statements in a manner acceptable to the translator for the language being used.

In the process, of doing this it can, and should, do the rule consolidation, optimizing, consistency checking, redundancy elimination, and so on that we have described in the earlier chapters of this book. It should also optimize action sequences, something we have *not* discussed.

The customary manner of describing the table to a preprocessor is by some form of punched-card input. The resulting output is a high-level language program that can either be punched into a card deck, stored on tape or disk for future use, or submitted directly to the appropriate compiler (high-level language translator).

C. AN INTERACTIVE DECISION-TABLE TRANSLATOR

Figures 9.1, 9.2, and 9.3 illustrate the workings of a decision-table translator written to demonstrate the process described in the preceding section. The entry of descriptive information and the control of the processor is *interactive*—that is, it is in the form of a conversation between the person constructing the table and the computer programs that will take the information he provides, store it, and, on demand, display it either in decision-table form or as a written procedure description. The exhibits are copied from actual typed inputs and outputs produced by a 2741 connected to a time-shared computer. The decision-table translator programs are written in APL, a high-level programming language.

Figure 9.1 Figure 9.1 illustrates the input required to describe our old friend the vintage-car decision table. The typed lines whose first character is an asterisk were typed by the computer; all the others were typed by the author.

ENTERTAB, the first entry, is the name of the APL program that accepts and stores a decision-table description. The program's first action is to request a name for the table. The answer VCAR will then identify the subsequent entries by name for later use.

The computer line C, A, R, EDIT requests an identification of the type of entry to be made. The first response, C, indicates that condition descriptions are to follow. The computer responds to this with ENTER CONDITIONS. Anything typed from this point on will be treated as a condition description until a blank line is entered by depression of the carrier key at the beginning of a line.

The format of permissible entries is indicated by the input examples in figure 9.1. If improperly formated entries are made, the program rejects them and waits for better input.

Actions are described and entered in the same way as conditions. Rules, however, are treated differently. The translator accepts rules in coded form. To assist the person entering rule information (and to verify the accuracy of the currently stored decision-table description), the program types out a codebook when the response R is made to the C, A, R, EDIT request. After the codebook has been typed, the computer types ENTER RULES. The entry of rules is terminated by a blank line. The entire input process is terminated by the entry of a blank line in response to the C, A, R, EDIT request from the computer.

Use of the EDIT option is not illustrated. It is the option used to correct or change an existing table description.

Figure 9.2 DISPLAYTAB is the name of an APL program that takes descriptive decision-table information and displays it in coded decision-table form. It is the only line in figure 9.2 that was typed by the author.

Note that the Duesenberg rule was entered in figure 9.1 as 3X:352. In figure 9.2, this is expanded into its three constituent simple rules.

Figure 9.3 If the condition and action entries made in figure 9.1 had been COBOL or FORTRAN statements, the translator would produce a COBOL or FORTRAN program as output. The translator identified as PROCEDURETAB in figure 9.3 does the same kind of translation but produces not a program but a procedure description in English. As in figure 9.2, only the first line, PROCEDURETAB 60, was typed by the author. The remainder was typed by the computer. The 60 portion of the first line was used by the PROCEDURETAB program to control the width of the printed line.

Note that the translator consolidates rules where possible. Parts 1, 3, and 5 of the procedure description all correspond to composite rules.

D. SUMMARY: DECISION-TABLE DATA BASES

An excellent discussion of the details of all aspects of decision-table translation is contained in a paper by H. J. Myers, "Compiling Optimized Code from Decision Tables," *IBM Journal of Research and Development* (vol. 16, no. 5, September 1972). Mr. Myer's paper is particularly useful, since it discusses how to optimize action sequences as well as condition sequences, a subject that is usually neglected. It also contains numerous examples (in flowchart form) of the output of his translator. The reader interested in comparing his hand optimizations against those produced by a computer program will find it a useful source of examples.

Up to this point, we have considered the decision table primarily as a stylized, formal method of procedure description. It can serve the purposes of nonprocedural description as well. Two developments make this point worth stressing at this time:

1. The growing use of information organized into generalized "data bases."
2. The growing use of "data communication" facilities to access data bases.

A data base is a pooling together of information to serve a variety of uses. Its physical organization on various storage devices is ordinarily not determined by any single use or "application" but by a variety of considerations. The "logical" description used by an application program will reflect only those aspects of data interrelationships important to the particular program. It will have little or no relation either to the "logical" structure used by other applications or to the "physical" structure on the disks or other storage devices. Describing these structures and transforming one description into another is an arduous, intricate task; so is understanding the descriptions once they have been written, and maintaining the descriptions when they have to be updated.

The interactive approach to decision tables we have discussed in earlier chapters and illustrated, in a small way, in this one can avail itself of the "data communications" aspect of most data-base programming systems. Structural relationships can be described concisely and vividly in decision-table terms. These descriptions can be generated, maintained, and queried by the kind of decision-table system described in chapter 3, section D, "Systems Analysis: Mechanizing the Paper-shuffling."

```
        ENTERTAB
*NAME?
VCAR
*C,A,R,EDIT
C
*ENTER CONDITIONS
MAKE IS (CORD, REO, DUESENBERG)
CONDITION IS (RUNNING WELL, RUNNING POORLY, NOT RUNNING)

*C,A,R,EDIT
A
*ENTER ACTIONS
COMMISSION IS (5 PCT., 10 PCT., VARIABLE)
SHOP TIME IS (NOT REQUIRED, ONE WEEK, TWO WEEKS, SIX WEEKS,
                            TO BE ESTIMATED)
MANAGER O.K. IS (NOT REQUIRED, REQUIRED)

*C,A,R,EDIT
R
*CODEBOOK
*MAKE IS          1:CORD
*                 2:REO
*                 3:DUESENBERG
*CONDITION IS     1:RUNNING WELL
*                 2:RUNNING POORLY
*                 3:NOT RUNNING
*=====================================
*COMMISSION IS    1:5 PCT.
*                 2:10 PCT.
*                 3:VARIABLE
*SHOP TIME IS     1:NOT REQUIRED
*                 2:ONE WEEK
*                 3:TWO WEEKS
*                 4:SIX WEEKS
*                 5:TO BE ESTIMATED
*MANAGER O.K. IS  1:NOT REQUIRED
*                 2:REQUIRED
*ENTER RULES
1 1:1 1 1
1 2:2 2 1
1 3:2 4 1
2 1:1 1 1
2 2:2 3 1
2 3:2 4 1
3 X:3 5 2

*C,A,R,EDIT
```

FIGURE 9.1   Interactive decision-table translator: Entering a table

```
        DISPLAYTAB
CODEBOOK

MAKE IS              1:CORD
                     2:REO
                     3:DUESENBERG
CONDITION IS         1:RUNNING WELL
                     2:RUNNING POORLY
                     3:NOT RUNNING
=========================================
COMMISSION IS        1:5 PCT.
                     2:10 PCT.
                     3:VARIABLE
SHOP TIME IS         1:NOT REQUIRED
                     2:ONE WEEK
                     3:TWO WEEKS
                     4:SIX WEEKS
                     5:TO BE ESTIMATED
MANAGER O.K. IS      1:NOT REQUIRED
                     2:REQUIRED

TABLE
```

MAKE IS	1	1	1	2	2	2	3	3	3
CONDITION IS	1	2	3	1	2	3	1	2	3
COMMISSION IS	1	2	2	1	2	2	3	3	3
SHOP TIME IS	1	2	3	1	3	4	5	5	5
MANAGER O.K. IS	1	1	1	1	1	1	2	2	2

FIGURE 9.2 Interactive decision-table translator: Displaying the table

PROCEDURETAB 60

1. IF MAKE IS CORD OR REO AND CONDITION IS RUNNING WELL: THEN COMMISSION IS 5 PCT., SHOP TIME IS NOT REQUIRED, MANAGER O.K. IS NOT REQUIRED

2. IF MAKE IS CORD AND CONDITION IS RUNNING POORLY: THEN COMMISSION IS 10 PCT., SHOP TIME IS ONE WEEK, MANAGER O.K. IS NOT REQUIRED

3. IF MAKE IS CORD OR REO AND CONDITION IS NOT RUNNING: THEN COMMISSION IS 10 PCT., SHOP TIME IS SIX WEEKS, MANAGER O.K. IS NOT REQUIRED

4. IF MAKE IS REO AND CONDITION IS RUNNING POORLY: THEN COMMISSION IS 10 PCT., SHOP TIME IS TWO WEEKS, MANAGER O.K. IS NOT REQUIRED

5. IF MAKE IS DUESENBERG: THEN COMMISSION IS VARIABLE, SHOP TIME IS TO BE ESTIMATED, MANAGER O.K. IS REQUIRED

FIGURE 9.3 Interactive decision-table translator: Translating the table into a procedure description

10

Open Questions and Opinions

A. DECISION TABLES: END OR BEGINNING?

We can regard decision tables in their present form as being either the final product of development activities that started in the late '50s or as a promising start for future development activities.

If we take the former view, there is little more to say. Competitive software suppliers will improve the performance of their preprocessors and extend the size of the tables they can handle. They will modify their offerings by adding a wrinkle here and a dimple there. But the decision table will be cast in concrete in its present form.

If we take the latter view, however, decision tables in their present form represent merely a promising beginning for development activities with great potential for systems analysis and programming.

As is probably apparent by now, I hold the latter view. Decision tables, as we now understand them, are a good idea. They can be used effectively in many areas. Unfortunately, the fact is that they are not now used in many places or by many people. In part, this is because of inertia. Accepted techniques are hard to replace. But in larger part the rejection of decision tables by systems analysts and programmers is based on the limitations of the standard form of decision table and of the translators based upon it.

In my experience, the people who have used the decision-table idea most effectively used neither the standard form nor a translator of any kind. They had complex interrelationships to describe and they used the two key decision-table ideas—the separate listing of conditions and actions, and the systematic checking of all possible condition sets—in their description. But in addition they used ad hoc conventions peculiar to the situation they were describing. In this latter kind of use, the history of decision tables is as old as the history of systematic thinking. The only thing that is new about the decision-table concept is the idea of using a digital computer to mechanize those parts of the descriptive process that can most effectively be mechanized and to convert the final procedure description into a computer program.

This use of the central ideas but rejection of existing forms and implementations suggests how great the need for research and development is. Virtually every systems analyst or programmer is confronted, and frequently confounded, by the problems of "complicated logic." How do the good ones tackle these problems? What techniques or conventions do they find useful? What mechanical aids do they feel it would be useful to have?

These are research and development questions that should be of interest to a variety of academic departments: computer science, industrial engineering, business—in fact, any department that uses computers for research or education.

Nor should this interest be confined to the academy. The key problems facing most modern organizations are those characterized by the use of the word *systems* somewhere in their description. These problems are growing in scope and importance. They cannot be—and are not being—solved efficiently by present techniques. New techniques should be developed. In their present form, decision tables are a new technique well worth using, studying, and developing.

The actual forms that future decision-table development activities take will, of course, be varied. Different avenues will be explored and the results of the exploration will determine what further avenues to pursue. It would be presumptuous of anyone to attempt to foretell what the future will reveal, but guidelines based upon a few of the questions currently open for investigation will perhaps suggest ideas.

B. OPEN QUESTIONS

1. *Using "or" rather than "and" to connect entry squares* Within a rule, the condition squares and action squares are connected by the logical function "and": "If a and b and c, then do d and e and f." It would be useful to have a way to connect all or some of these squares by "or." There is at present no standard way to do this. One that is both intelligible and easy to apply would help avoid some of the awkwardness and multiplicity of rules that characterize decision tables with exclusively "and" connections between squares.

2. *Generalizing the nature of the information in the entry squares* We have already done this in the case of actions squares by entering a sequence number where it was needed (figure 8.8). It could, in many cases, be

useful to enter other kinds of information as well: values for functions to be evaluated; the names of functions to be used in a general functional expression; the relative frequency of each of a group of actions that occur randomly in a simulation program; and so on. For example, we might have a billing program in which the tax to be applied might vary from state to state. The action in the action stub might be "apply tax X." The X value to be used could be provided in the action entry. Or it might be more general: "Apply tax from appropriate tariff." Then the action entry would be the name of the table in which that tariff was stored.

3. *"Parameterizing" the decision table* Many tables might have an identical structure, as far as the meaning and number of the rows and columns that make them up, but differ in the actual values of the conditions and actions. It would save a great deal of time, effort, confusion, and bookkeeping if a single "prototype" table could be stored that could be used to produce a specific working table. This could be done by a program that would receive, as input, a list of parameters giving the actual values to be used and the name of the decision table that described the structure desired.

4. *Decision-table libraries* Decision-table libraries should be kept available for use within a computing system on the same basis as other program libraries. But they have a structure that is different from that of conventional programs. How can a decision-table library best be organized?

5. *Using branch labels in debugging* We described earlier how a sequence of branch labels could give us the branching history of a program. Such labels could automatically be generated and stored but most of them would only be of interest during the "checkout" or "debugging" part of a program's life. The conditional assembly features of most assemblers could be used to cause storing of these branch labels when a program is being tested. After testing is completed, a global variable could be changed from "Test" state to "Production" state; a reassembly or recompiling of the program would then omit the instructions that store the branch labels. This variable could, of course, be changed back to "Test" status when the program is modified or when errors occur in its operation.

6. *Using peripherals effectively* The typewriter or typewriterlike terminal is currently the most widely used device for communicating with a time-sharing computer. Other devices are in use, however, and their use is spreading. Some of these have great implications for the "interactive" kind of decision-table processing we have discussed and illustrated. The cathode-ray tube, for example, provides an extremely effective way to display structural relations whose display would be time-consuming and awkward on a typewriter. This is just one among many peripheral devices whose existence should be recognized in decision-table development activities.

7. *"Top-down" programming* As programming systems grow larger and more comprehensive, the sizes of the programming groups required to produce them increase at an even faster rate. The tasks of managing such groups—planning, coordinating, and controlling their activities—become extremely difficult as the groups grow in size.

Various techniques have been proposed to assist in the programming-management process. One of them is "top-down" programming. In top-down programming, the first step is to divide the planned system into components. These are described as programs whose externals are known but whose internals have yet to be designed. The external features of a program are its inputs, its outputs, and its connections either to other programs or to input and output from outside the system. Dummy programs to accept typical inputs, generate typical outputs, and test interconnections can be written to check out and describe this highest-level design.

Each of these high-level building-block programs can, in turn, be subdivided into smaller blocks in the same way. This second-level description can be further subdivided, and so on, until the moment of truth when the designing stops and the programming begins. At this point, we have various levels of description. At any level, the interconnection between subsets of components describes the inner workings of the components at a higher level.

The structure of decision tables could serve the purposes of top-down programming extremely well. The only extensions needed to the basic decision-table idea would be (1) level identification; (2) an identification of the table's place in the higher-level description; (3) an identification of the lower-level tables into which it is subdivided.

C. DECISION TABLES: AN INTRODUCTION

The open questions discussed briefly in section B are merely a few of the many that spring to mind when decision tables are viewed as a beginning rather than an end. This book will have served its purpose if it persuades some of its readers that decision tables are a promising starting point for the programming, systems analysis, design, and documentation of the future rather than a perfected technique for doing the programming, systems analysis, design, and documentation of the past.

A question frequently asked about decision tables by people who believe in their merits is: "Why aren't they used more widely?" This question was discussed briefly in the opening section of this chapter. The best way to end a book that is truly intended as an introduction rather than as the final word is to return to that key question and express some highly personal, highly individual opinions.

I believe decision tables have not been more widely accepted, used, and developed because—

1. The main historical emphasis has been on limited-entry rather than extended-entry decision tables.
2. Decision tables have been thought of primarily as a programming-language supplement rather than as formal tools for a much more inclusive set of important problems in both systems analysis and programming.
3. The "human engineering" aspects of decision tables have been neglected.

Limited-entry Tables

Limited-entry tables are bad ways to describe any job that is not inherently yes-no in character. They are bad for two important reasons—

1. They replace the natural questions and actions that actually occur in a procedure by unnatural, derived questions and actions that obscure the procedure's fundamental structure. (Compare the extended-entry and limited-entry forms of the vintage-car decision table.)
2. They require extremely large tables for extremely simple problems, thus generating both incomprehensibility and a greatly increased likelihood of coding errors.

When a simple question like "What is the make of the car?" has to be replaced by a set of as many yes-no questions as there are types of car on a used-car lot, a procedure description veers rapidly toward a tedious kind of incomprehensibility. Not only that, the number of rows in the table itself will change as new makes are added or dropped. This kind of change is a good deal more drastic than adding or dropping rules, which is all that would have to be done to modify an extended-entry table. The limited-entry versions of most clerical procedures are so unnatural that they might very well discourage people about the whole decision-table idea.

In a sense, the exponents of limited-entry tables reveal a peculiar kind of flowchart bias. Most flowcharts are drawn with binary branches. This is unfortunate, since even in flowcharts there are many more efficient ways to indicate branching. Within the computer itself, the "natural" form of branching is, for most existing computers, not two-way but three-way. For *any* computer, the most efficient way to branch to one of many possible paths is most likely to be by branching unconditionally to a calculated address. Thus there is no practical reason to prefer two-way branching over alternative forms and there are a great many reasons to prefer alternative forms in many cases.

The vintage-car written procedure, extended-entry table, and limited-entry table each have nine valid simple rules. The limited-entry table, however, also has fifty-five *in*valid simple rules. These have to be caught by some kind of error-detection programs. This trivial example is only a small indication of the burden imposed on a procedure when a limited-entry table is used for an extended-entry job. When two straightforward questions are replaced by six roundabout ones, it is not only comprehensibility that suffers, it is efficiency as well. The limited-entry table is not only a poor procedure description; it encourages, in fact almost compels, bad programming practices.

Programming and Systems Analysis

The chief difficulty a programmer faces is not the intrinsic difficulty of programming itself. He regards the challenges of doing the best programming job possible as being the very things that make his job worthwhile. They give him the opportunity for creative achievements that are among the most important rewards of the profession he has chosen.

His major difficulty—and one that provides no satisfactions to counterbalance the frustrations it brings—is his inability to determine just what it is he is supposed to program. The instructions he is given are usually incomplete, unintelligible, and contradictory. Having, after much wrangling, determined what they are most likely to mean, he is very commonly told that a "new approach" has been taken and that he is to disregard all he has previously been told.

The difficulty, of course, is not in the programming portion of the system development process. It is in the "systems analysis" portion. The programmer is at the end of a long chain of interactions which will, eventually, produce a working computer procedure. Until the buck is passed to him, nonsense can parade as wisdom, blue sky can masquerade as good green earth. But before the programmer completes his task, the nebulosities must be solidified, the abstract generalizations must be transformed into specific computational actions, the glowing promises of other people must be redeemed by the program that has been assigned to him as a relatively insignificant part of the "big picture" that others have painted.

Given the way he gets his assignment, of course his task is difficult. But it shouldn't be. Before a programmer writes his first instruction, it should be completely, definitely, clearly established what his program is supposed to do.

In practice, this is not usually the case. The usual case is the chaotic one we have pictured above. This does, indeed, make for a difficult, almost impossible, job. But this is not the way it should be. Programming specifications—system design—should be determined *before* a programmer starts to work, not after.

Yet the mechanical or conceptual aids that have been developed to assist in mechanizing procedures have focused on programming as the chief bottleneck. Decision tables have, in most of the literature, been treated primarily as an aid to programmers or as a "high-level language" that will help make programming easier. This is the primary motivation for developing "preprocessors"—computer programs to superimpose a new programming aid (decision tables) on an old one (high-level languages).

Laudable as these efforts to help the programmer are, they miss the point. The best way to ease the programming bottleneck is to prepare better programming specifications. This is where decision tables show their brightest promise. It is also where, at present, the least has been done to redeem that promise.

Human Engineering

This is one of the ugly neologisms that characterize the modern flood of effusions about *automation, cybernation, systems analysis, modular, interactive, synergistic, symbiotic man-machine communication,* and so on. The new terms must be used, because most people in the field have grown accustomed to them. But anyone interested in precise communication heaves a sigh when he encounters one or is forced to use one and thinks wistfully of the days when the development of technical terminology was more orderly, more thoughtful, more rational, and more scholarly. Wouldn't it be delightful to restore to technical writing the kind of thoughtfulness and care that existed in the days when Fowler could look over the physicist's shoulder and chide him gently, not for creating a new technical term like *impedance,* but for failing to recognize that a proper derivation from its Latin origins would have been *impedience?*

In any event, *human engineering* is the term we have been given by the human engineers. (The inhuman engineers have not been heard from.) By the "human engineering" aspects of a system we mean those features that affect its ease of use by the people who will have to use it.

"Human engineering" is neglected throughout the fields of systems analysis and programming. We have made the point throughout this book that systems analysis is neither systematic nor analytic. Its plight is that of the barefoot children of the shoemaker. It perfects, eases, systematizes, regulates the procedures of other people, but its own procedures are either nonexistent or cumbersome.

Coding sheets are an example of the lack of imagination that characterizes the procedures of systems analysis and programming. In preparing a program, it is mechanical requirements, not human requirements, that are reflected in the program-preparation process. A string of symbols, properly punctuated, must be carefully placed in a succession of little spaces—just the *right* little spaces. The letter *O* and the numeral 0 must be carefully distinguished by a line through one or the other; the letter *I* must be distinguished from the numeral 1 by a line under either one or the other, and so on for 2 and *Z* and any other characters that are likely to be misread by the key-punch operator. The permissible characters are those that are available on the particular key-punch in use at the time; the permissible combinations and sequences of characters are those required by the compiler, link editor, loader, and so on. The whole process is a burdensome one in which the inflexible characteristics of machines or mechanical procedures determine what has to be done and how it has to be done.

The ordinary form of decision table for "preprocessor" use is merely the coding sheet writ large. The fundamental restrictions are the same, and new ones are added. If there is to be any creative, analytic, design use of decision tables, it must be done elsewhere, in some other, unspecified manner. Like most of the really important activities of systems analysis and programming, few aids are available for the creative activity of developing decision tables in the first place. What should be a liberating idea becomes a straitjacket of mechanical restraints.

The point is an important one and therefore a good one with which to conclude. The truly important idea underlying the use of decision tables is the efficient description of complicated relationships as a correspondence between *state* vectors and *action* vectors. For the nonmathematician, these are formidable terms. They shouldn't be. A state vector is a set of coded condition entries; an action vector is a set of coded action entries. The mathematical jargon describes an extremely practical reality: the need for codebooks to describe any kind of procedure we wish to systematize. Decision tables are a way to develop, maintain, and use such codebooks in an efficient and productive manner. As things now stand, our most pressing need is for decision-table aids that will help us achieve this efficiency and productivity.

BIBLIOGRAPHY

Books

London, Keith R., *Decision Tables*, Auerbach, 1972.

Pollack, Solomon R.; Hicks, Harry T., Jr.; Harrison, William J., *Decision Tables: Theory and Practice*, Wiley, 1971.

Both these books contain decision-table bibliographies, the latter a particularly comprehensive one.

Papers

Myers, H. J., "Compiling Optimized Code from Decision Tables," *IBM Journal of Research and Development,* vol. 16, no. 5, September 1972.

Montalbano, M. S., "Tables, Flow Charts and Program Logic," *IBM Systems Journal*, vol. 1, no. 1, September 1962.

The Myers paper contains many key ideas important in decision-table translation. The Montalbano paper is an early discussion of ideas developed more fully in this book.

Answers to Exercises

Chapter 2

In the first four exercises, there are three sources of information about the nature of each figure:

    a. The placement of the double lines
    b. The occurrence of exactly two or of more than two entries along any row
    c. The actual entries along a row

1. action entry of an extended-entry table
2. condition stub of a limited-entry table
3. action stub of an extended-entry table
4. rule of an extended-entry table
5. See procedure description in section 2.B.1.
6. See procedure description in figure 3.1.
7. See flowcharts in figures 3.2 and 3.3 and chapter 8.
8.

Header	1	2	3	Else
Sick pay duration (days)	1–7	8–30	Over 30	
Sick pay rate (pct.)	75 or less	75 or less		
Hospitalized?	Yes			
Read chapter 16	X	X	X	
Do not qualify for sick pay exclusion				X

The fourth rule in this table, identified by the word *Else*, is our first instance of a catchall kind of rule, the kind that applies when none of the others do. The customary convention is not to number the rule but to identify it, as we have done, with the heading "Else," and to make no entries in the condition portion of the rule. *Else* rules are discussed and analyzed in section 6.D.

9. a.

	1	2	3	4	5	6	7	8
Outfielder?	N	N	N	N	Y	Y	Y	Y
Left-hander?	N	N	Y	Y	N	N	Y	Y
Rookie?	N	Y	N	Y	N	Y	N	Y
Redhead	N	N	N	N	Y	Y	Y	Y
Glasses	N	N	Y	Y	N	N	Y	Y
Over 40	N	Y	N	Y	N	Y	N	Y

b. A veteran right-handed pitcher prefers girls who are not redheads, do not wear glasses, and are not over forty.

10. The difference between the solution diagram (identified as the "House-pet-owner diagram" on the next page) and a true decision table is fundamentally the difference between the ways in which we use expressions of the form "if . . . then."

In a decision table, the "if . . . then" is a conditional directive. It is shorthand for the command "If X is the case, then do Y," where X is a set of conditions and Y is a set of actions.

In the diagram, the "if . . . then" is the "if . . . then" of logic. It is shorthand for "If X is true, then Y must be true."

	1	2	3	4	5
COLOR: RWB	B	R	~W		
PET: OFA	F	~O		F	~A
OWNER: TDH		T	D	H	
R O T		~P			
R O D		~P		~P	
R O H		~P			
R F T	~P			~P	
R F D	~P				
R F H	~P			~P	
R A T						Tom lives in the red house and owns an alligator.
R A D				~P	
R A H					~P
W O T			~P		
W O D				~P	
W O H						Harry lives in the white house and owns an otter.
W F T	~P		~P	~P	
W F D	~P				
W F H	~P			~P	
W A T			~P		
W A D				~P	
W A H					~P
B O T	~P				
B O D	~P			~P	
B O H	~P				
B F T				~P	
B F D						Dick lives in the blue house and owns a fox.
B F H				~P	
B A T	~P				
B A D	~P			~P	
B A H	~P				~P

House-pet-owner diagram

Consider rule 1 in the diagram. It corresponds to the first statement in the puzzle, though the resemblance is probably not immediately apparent. Puzzles of this kind conceal the essential information they convey by mixing it in with as much nonessential information as possible. If we strip the nonessential information from statement 1 of the puzzle, we get the essential information: the pet of the man who owns the blue house is a fox.

Now, from the conditions of our problem, we know that if this is true then the following statements are true as well:

a. The owner of the red house does not keep a fox as a pet.
b. The owner of the white house does not keep a fox as a pet.
c. The owner of the blue house does not keep an otter as a pet.
d. The owner of the blue house does not keep an alligator as a pet.

The "actions" listed in the action stub of the diagram are combinations of house colors, pets, and owners. We code each possible value of these by using its initial letter to represent it. We then combine these codes in all possible ways. There are 27 combinations in all and we see them in the action stub of the diagram.

We make only one kind of action entry, "∼P," for "Not possible." Our strategy is that of Sherlock Holmes, to eliminate all possibilities until only the true one is left.

The rule we have just discussed (rule 1) eliminates twelve possibilities. (See the diagram and compare rule 1 with the discussion above.) The second rule eliminates three more. The third also eliminates three, but only two of these are new; the other has been previously eliminated. The fourth eliminates twelve, of which five are new. The fifth eliminates three, of which two are new. This leaves us with three possibilities. Since three houses, pets, and owners are in question, these three represent the only possible matching that satisfies all the requirements.

In solving the problem, we stripped from each of the five rules any information that was not relevant to house color, pet, or name of owner. Once the problem is solved, however, we can return to the original information and deduce some surprising consequences of the knowledge we have so painfully acquired; for example that Tom, although he lives next door to his brother-in-law and shares with him an antipathy for foxes, never goes into his house.

11. The dependent rows of figure 2.2 can be grouped as follows:

Make

    Is car a Cord?
    Is car a Reo?
    Is car a Duesenberg?

Condition
> Is car running well?
> Is car running poorly?
> Is car not running?

Commission
> Commission is 5 percent.
> Commission is 10 percent.
> Commission is variable.

Shopwork
> No shopwork needed.
> Schedule 1 week of shopwork.
> Schedule 2 weeks of shopwork.
> Schedule 6 weeks of shopwork.
> Estimate shopwork.

Note the relation between the interrow dependency in the limited-entry table and the basic conditions and actions of the extended-entry table for the same procedure.

Chapter 3

1. This example has four major objectives:
   a. To show how language changes in the course of a relatively short time
   b. To show how inadequately technical terms or jargon are described in dictionaries for general use
   c. To illustrate the explosive, inconsistent growth of jargon in the literature of data processing
   d. To illustrate how much of the meaning of a sentence depends on the meanings of groups of words rather than the meanings of the individual words themselves.

   Changes in language (*Garble, comprise, verbal, infer, Congressman.*) *Garble* and *Congressman* are discussed below. *Comprise* is, or used to be, a useful word that was the inverse of *compose*. In much current technical writing it is used to mean exactly the same as *compose*. *Infer* is inverse to *imply*. In much speech and writing, it is used where *imply* should be used. *Verbal* is frequently used to mean *oral*. Most companies have forms with a heading like: "Avoid Verbal Orders." Literal obedience would mean transmitting orders by sign language, smoke signals, pictures, diagrams, or some other way that did not use words. This is not the intent of the form; its intent is to encourage managers and supervisors not to avoid words but to write them rather than say them.

Technical jargon (Cobble, coil.) Precise communication in specialized fields requires the careful use of narrowly defined technical terms. Sometimes these are words that have different meanings in ordinary speech; sailors, for example, use *sheet* and *bend* in senses different from the ordinary. *Cobble* and *coil* are instances of the jargon of the steel industry. A *cobble* is a piece of finished or semifinished steel that has been damaged in manufacture. A *coil* is a long steel sheet that has been wrapped around a spindle. Its shape is *not* that of a helix, the shape that the word coil suggests in most uses.

Data-processing jargon (File, compile, assemble, processor.) The use of these words in data processing is not only inconsistent with their ordinary meanings; it is sometimes not even consistent within the field. As data-processing jargon, *file* and *processor* have both hardware and software meanings. A *file* can be either a physical device like the IBM 2314 or IBM 3330 or a conceptual collection of "records" on a tape, disk, drum, or other device. Within the latter use, it can again have two meanings, ordinarily differentiated as *logical file* and *physical file*. And even this last refinement is currently being further confused by the growth of "data base" systems in which the physical file represents an intricately connected network of "logical" files. *Processor* can mean either a particular computer model or a language translator. *Compile* and *assemble* refer to acts of translation rather than to any acts suggested by their ordinary meanings.

The meaning of individual words (UP). Much of the life of a language is contained in its smallest words—its *particles*. Anyone who has ever looked up *noch, nach*, or *da* in a German dictionary will recall the frustration of confronting the number of choices he has to consider when he attempts to translate a sentence in which one of them appears. In this respect, English is, if anything, even worse. Some of the possible uses of the particle *up* are illustrated in question 3.

2. The Ems Dispatch, published by itself, without any explanation of the background out of which it had arisen, what other discussions had been taking place between France and the Kaiser, what German points had been raised or pressed—in other words, without an explanatory context—seemed to be an insulting ultimatum from France to Germany. It contributed to bringing about the Franco-Prussian war, Prussian dominance over a united Germany and, as eventual consequences, World Wars I and II.

The original, useful, meaning of *garble* was "bias a statement by including only the facts favorable to one point of view." This meaning is virtually lost. This is unfortunate, since electronic technology has made garbling, in the original sense, a terrifyingly easy, extremely effective way to present a distorted picture of reality. A politician can be goaded to the point of frenzy by an antagonistic interviewer seeking a sensational story. A skillful videotape editor, selecting one of the politician's less cautious answers without providing a context for it, can influence millions of viewers to think of him as a red-faced, clenched-fist extremist willfully uttering irresponsible views. In its original sense, the word *garble* described such an action precisely; now that this meaning has been lost, we have no single word that will do as well.

3. I could not do this exercise.

4. If you are a citizen of one of the fifty states, the answer is 3: two senators and a representative. I am not sure of the answer for citizens of the District of Columbia, Puerto Rico, territories, mandates, insular possessions, and the like.

   Even the most respected news publications speak of congressional elections as "races for seats in Congress and the Senate," making it sound as if the Senate were not part of Congress. This usage is unfortunate, since it influences the way voters think about the relative importance of senators and representatives, causing them to forget, for example, that the man next in line for the presidency, after the vice president, is the speaker of the House of Representatives.

5. As written, the sentence probably can no longer be read the way it was intended. Yet it describes the fundamental characteristic that differentiates a stored-program computer from all preceding calculating devices. In a stored-program computer, the form of the data is identical with the form of the instructions used to process it. In other words, the instructions themselves are a form of "data." The power of the stored-program computer derives precisely from this ability to have instruction "data" process other forms of data. But virtually all readers of computer literature now read *data processing* as one word (the hyphen in *data-processing* is usually omitted even when the combination is used as an adjective). This makes unintelligible any sentence that uses the words as they were used in the exercise question.

6. See the "Flowchart: Invoicing" and "Decision Table: Invoicing" illustrations. Compare the table with figure 7.12 in chapter 7.

7. See the "Flowchart: Payroll" and "Decision Table: Payroll" illustrations.
   The decision table, as it stands, is in "standard" form—the only form we have discussed up to this point. Other forms can be more informative and concise. A modified form that is useful for this example is discussed

in section 7.E. It avoids the repeated "0,1" entries along the "Seventh day?" condition row and generalizes the action-entry squares to show the order in which actions must be taken. A table incorporating these ideas is given in the answer to exercise 8.5.

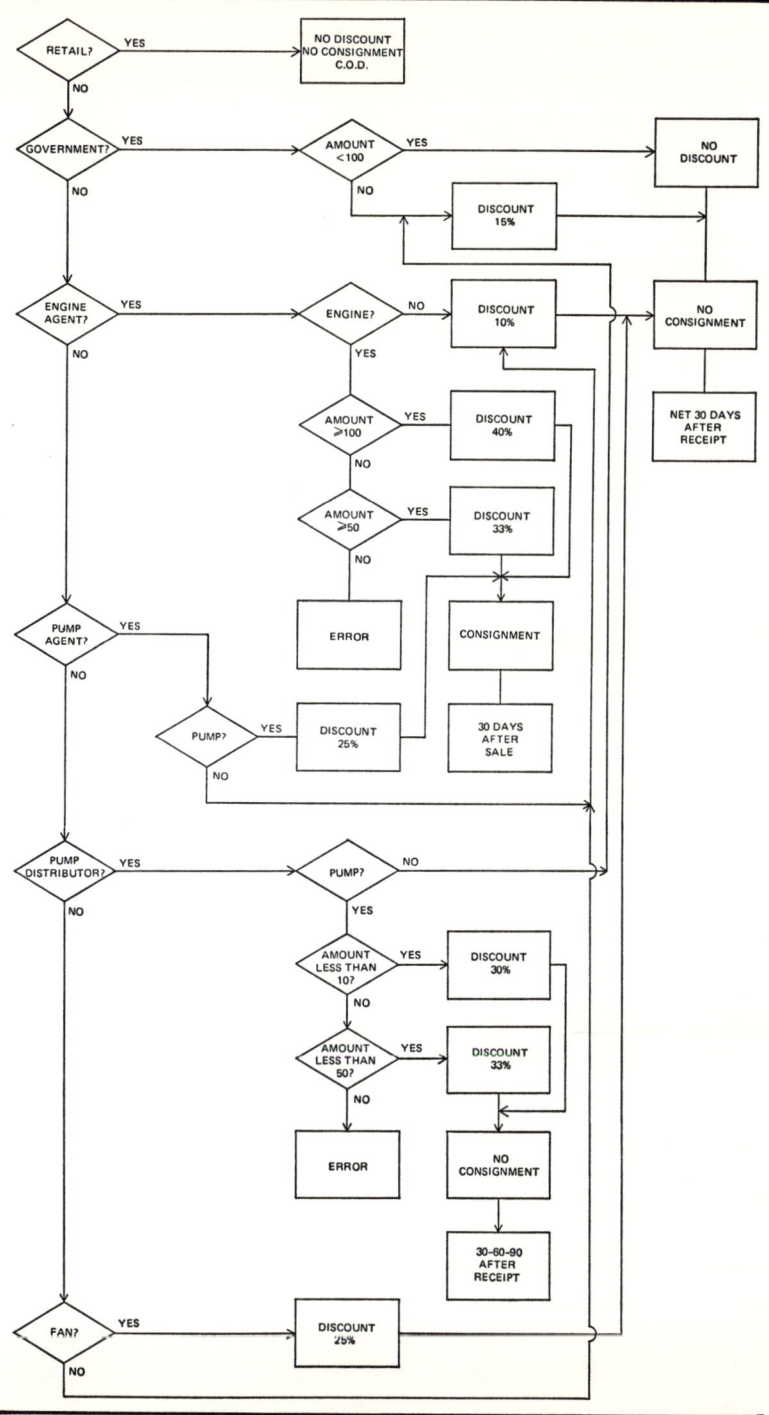

EXERCISE 3.6 Flowchart: Invoicing

CUSTOMER IS
   1:RETAIL
   2:GOVERNMENT
   3:ENGINE AGENT
   4:PUMP AGENT
   5:PUMP DISTRIBUTOR
   6:FAN DISTRIBUTOR

AMOUNT IS
   1:$A<10$
   2:$10 \leq A < 50$
   3:$50 \leq A < 100$
   4:$100 \leq A$

PRODUCT IS
   1:ENGINE
   2:PUMP
   3:FAN

DISCOUNT IS
   1:0
   2:10
   3:15
   4:25
   5:30
   6:33
   7:40

CONSIGNMENT IS
   1:NO
   2:YES

TERMS ARE
   1:C.O.D.
   2:30 AFTER RECEIPT
   3:30 AFTER SALE
   4:30-60-90 AFTER RECEIPT

	1	2	3	4	5	6	7	8	9	10	11	12	13	E
Customer	1	2	2	3	3	3	4	4	5	5	5	6	6	
Amount	X	~4	4	X	3	4	X	X	X	1	2	X	X	
Product	X	X	X	~1	1	1	~2	2	~2	2	2	~3	3	
Discount	1	1	3	2	6	7	2	4	3	5	6	2	4	
Consignment	1	1	1	1	2	2	1	2	1	1	1	1	1	
Terms	1	2	2	2	3	3	2	3	2	4	4	2	2	
Error														X

EXERCISE 3.6 Decision table: Invoicing

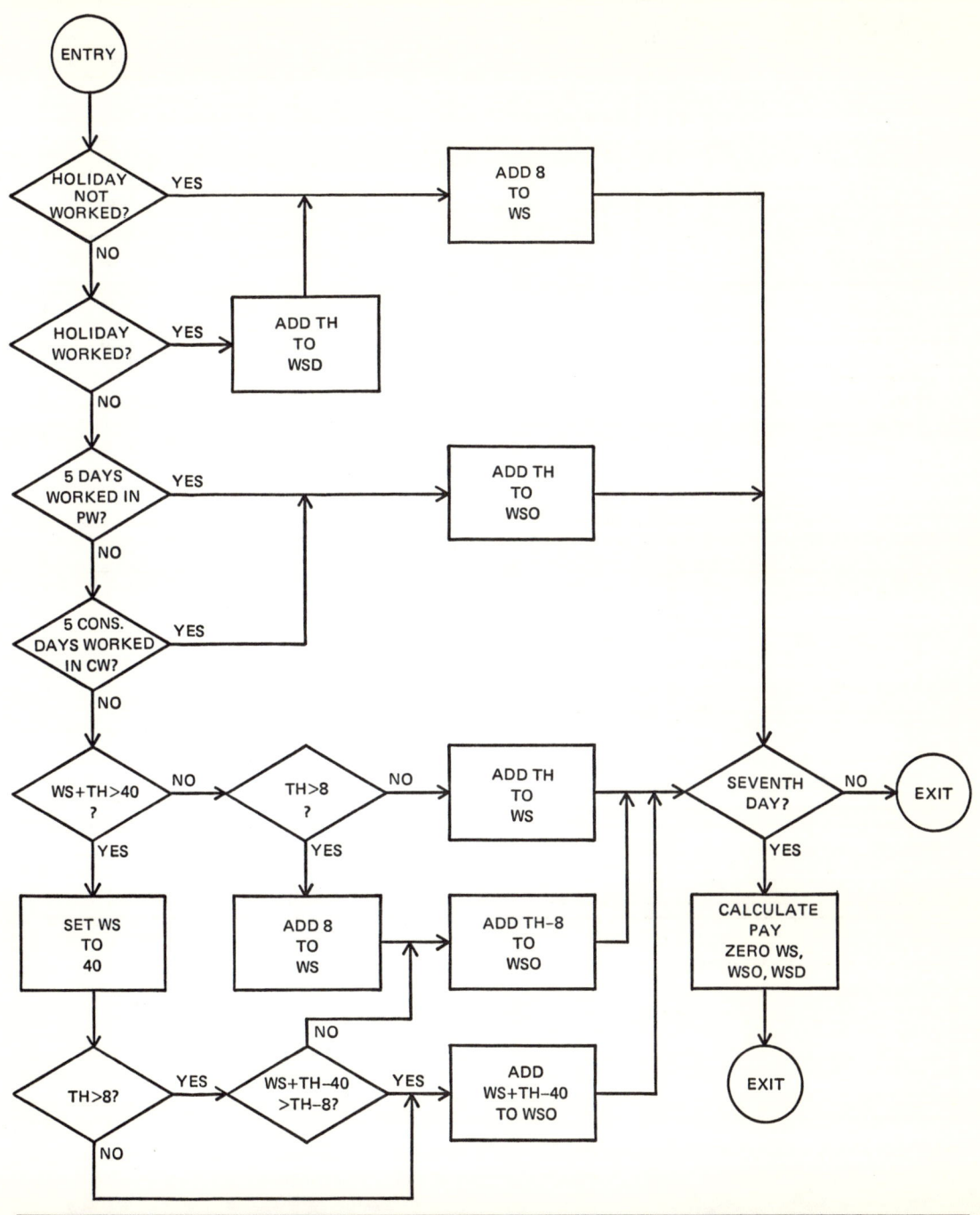

EXERCISE 3.7 Flowchart: Payroll

NW: Holiday not worked PW: Payroll week   WS: Straight-time hours
W: Holiday worked   CW: Consecutive 7 days WSO: Time-and-a-half hours
NH: Not a holiday   TH: Time-card hours   WSD: Double-time hours

	1	2	3	4	5	6	7	8	9	10	11	12	13	14	15	16	17	18
Holiday (NW, W, NH)	1	1	2	2	3	3	3	3	3	3	3	3	3	3	3	3	3	3
5 PW days worked?					1	1	0	0	0	0	0	0	0	0	0	0	0	0
5 Cons CW days worked?							1	1	0	0	0	0	0	0	0	0	0	0
(WS+TH)>40?									0	0	0	0	1	1	1	1	1	1
TH>8?									0	0	1	1	0	0	1	1	1	1
(WS+TH−40)>TH−8?															0	0	1	1
Seventh day?	0	1	0	1	0	1	0	1	0	1	0	1	0	1	0	1	0	1
Add TH to WSD			X	X														
Add 8 to WS	X	X	X	X							X	X						
Add TH to WSO					X	X	X	X										
Add TH to WS									X	X								
Add TH−8 to WSO											X	X			X	X		
Add WS+TH−40 to WSO													X	X			X	X
Set WS to 40													X	X	X	X	X	X
Calculate pay		X		X		X		X		X		X		X		X		X
Zero WS, WSO, WSD		X		X		X		X		X		X		X		X		X

EXERCISE 3.7 Decision table: Payroll

Chapter 4

1.
1	1	2	3	4	4	5	6	6	7
1	1	1	1	1	2	1	1	2	2
1	2	2	2	2	3	4	4	3	3
A	B	C	D	E	F	G	H	I	J

2.
	X	X											
X	X	X	X				X	X					
			X	X									
				X	X								
					X	X				X	X		
								X	X				
								X	X	X	X		
	X		X		X		X		X		X		X
	X		X		X		X		X		X		X
A	B	C	D	E	F	G	H	I	J	K	L	M	N

3. EXERCISE 3.6   Rules 1, 2, 3, 4, 7, 8, 9, 12, 13, ELSE

   EXERCISE 3.7   Rules 1 through 14
   (See tables in answers for chapter 2)

4.
1	X	X	1	1	2	2	2	2	2	2
X	X	1	2	3	1	1	2,3	2	3	3
2	1	2	2	2	2	2	2	2	2	2
1,2	X	3	3	3	1	2	1	2	3	2	3
A	B	C	D	E	F	G	H	I	J	K	L

5.

1	1	1	1	1	1	1	1	1	1	1	1	1	1
1	1	1	1	1	1	1	1	1	1	1	1	1	1
0	0	0	0	0	0	0	0	0	0	0	0	0	0
0	0	0	0	0	0	0	0	0	0	0	0	0	0
0	0	0	0	0	0	0	1	1	1	1	1	1	1
0	0	0	0	1	1	1	1	0	0	0	1	1	1
0	0	1	1	0	0	1	1	0	0	1	1	0	0	1	1
1	1	1	1	1	1	1	1	1	1	1	1	1	1
0	1	0	1	0	1	0	1	0	1	0	1	0	1

6a.

1
2,3
2

6b.

1	2
1	1
2,3	2,3
2	2

6c.

1	2	1	2
1	1	1	1
2	2	3	3
2	2	2	2

6d.

1	2	1	2	1	2	1	2	1	2	1	2
1	1	1	1	1	1	1	1	1	1	1	1
2	2	3	3	2	2	3	3	2	2	3	3
2	2	2	2	2	2	2	2	2	2	2	2
1	1	1	2	2	2	2	3	3	3	3	

The rules are the same as those of figure 4.4 but in a different order.

7. The following pairs of rules overlap

$$(1,5) \ (1,6) \ 2,6) \ (4,6)$$

Chapter 5

1a. 86400

b. 36 (allowing for 0 feet 0 inches)

c. 5760 combinations less than a pound
   5760 grains in a pound

2. 130 if the vowels are a,e,i,o,u
   156 if the vowels are a,e,i,o,u,y
   (26x5) and (26x6) respectively

3. CAME CAMS CAPE CAPS CARE CARS COME
   COPE CORE COPS CURE CURS CUPS DAME
   DAMS DALE DARE DOLE DOME DOPE DUPE
   FAME FARE FOPS FORE FUME FURS GALE
   GAME GAMS GAPE GAPS GARS GORE GUMS

   61 combinations are not words.

4a. (BAD, BAR, BAN, CAB, CAD, CAN, CAR, FAD, FAN, FAR)
 b. (FAD, FAN, FAR, FEW, FUN)
 c. (BED, FEW)
 d. (FAD, FAN, FAR, FEW, FUN, BED)
 e. (FEW)
 f. (FAD, FAN, FAR)
 g. (ϕ)   (THE NULL SET)

5. OVERLAP (M,A,N)
   UNION   (M,E,A,N,Y)
     (NO PUN INTENDED)

6. The decision table is complete. No rules overlap.

7a. 64 rules are possible. Only 9 are shown. 55 are missing.

 b. An "Else" column should be added.
    It should lead to an error exit.

Chapter 6

1. Rules 1 and 5 overlap in rule 22131.
   Rules 1 and 6 overlap in rule 21131.
   Rules 2 and 6 overlap in rule 51131.
   Rules 4 and 6 overlap in rule 31131.

   Since rule 1, 2X131, consists of two simple rules, one of which is overlapped by rule 5 and the other by rule 6, it is completely overlapped.

2. As we see from the preceding answer, this table is inconsistent if rules 1, 5, and 6 do not all specify the same action. If they do, the overlap of rule 1 may be eliminated simply by dropping it.

   Rule 2, 511XX, overlaps rule 6 in rule 51131. Rule 2 contains twelve simple rules, of which we wish to eliminate the one that is overlapped. This can be done by replacing rule 2 by the pair of rules:

   511(~3)X covering nine simple rules
   5113(~1) covering two simple rules

   The overlap between rules 4 and 6 can be removed by replacing 4 by a pair of rules just like those shown except that they have a 3 in the first position instead of a 5.

3. 311, 321, 532, 542. These may be seen as the entries without corresponding action sets in the policy map of figure 7.14.

4. The overlap is rule 11(2,3)2(1,2). Rule 1, X1(2,3)2X, covers twelve simple rules of which the four in the overlap are to be removed. This can be done by replacing rule 1 by two rules, 21(2,3)2X, covering six simple rules, and 11(2,3)23 covering two simple rules. It would, of course, be simpler to remove the overlap from rule 2 by replacing it by 11(1,4)2(1,2). However, the choice is not arbitrary. Since different actions are taken by the rules that overlap, the overlapped rules must be assigned to one action or the other.

5. The second-row codes are 1, 2, 3, 4; the third-row codes are 1, 2, 3, 4, 5. The impossible combinations are 11, 13, 14, 15, 21, 23, 24, 25, 31, 32, 41, 42, 43, 44.

6. The table is given in the "Color-Width-Size" illustration. Only the entry portion is shown. The letters A and R are used for "Accept" and "Reject," respectively, instead of the codes shown in the codebook, to emphasize the distinction between the condition portion of the table and the action portion.

CODEBOOK

COLOR IS   1:61
              2:62
              3:51
              4:60
              5:54
              6:56
              7:55
              8:70
WIDTH IS   1:B
              2:C
              3:D
              4:E
              5:EE
SIZE IS   1:7
              2:7 1/2 TO 8 1/2
              3:9 TO 11
              4:11 1/2 TO 12
==============================
ACTION IS   1:ACCEPT
              2:REJECT

6,7,8	~(2,4)	X	6,7,8	1,3,5	ELSE
1	2	3	4	5	
3,4	3	2,3,4	1,2,3	2,3	
A	A	A	A	A	R

EXERCISE 6.6 Color-width-size table

Chapter 7

1. The table below describes the sixteen logical functions in terms of the simpler functions *false, true, and, or, not, equal, not equal.*

```
0011:A  (0,1)
0101:B  (0,1)
====:======
0000:IDENTICALLY FALSE
0001:A ∧ B
0010:A ∧ (~B)
0011:A
0100:(~A) ∧ B
0101:B
0110:A≠B (EXCLUSIVE OR)
0111:A ∨ B
1000:(~A) ∧ (~B)
1001:A=B
1010:~B
1011:A ∨ (~B)
1100:~A
1101:(~A) ∨ B
1110:(~A) ∨ (~B)
1111:IDENTICALLY TRUE
```

Note that, if all the binary digits in the function code are alike, the input values are irrelevant. If they are all zero, the output value is zero (or False) for all possible combinations of input values. If they are all one, the output value is always one (or True).

How can these possibilities arise? They can arise in the same way the simplified form $B' + C'D$ arose from the expression $AB'C'D' + B'C + A'B'C' + AC'D + A'BC'D$. The input expression can be a complicated one. It is not until it is simplified that we see that it is either identically true or identically false.

The functions that have a single one and three zeroes are all AND functions. They correspond to all possible combinations of one of the pair (A, A') with one of the pair (B, B').

The functions that have a single zero and three ones are all OR functions. They correspond to all possible combinations of one of the pair (A, A') with one of the pair (B, B').

The functions with two ones and two zeroes either depend on only one of the variables or depend on the equality or inequality of one to the other.

2. The table below describes the operations required to prove the theorem. It can be expressed in many forms.

$$
\begin{array}{rll}
(\sim X) & 1100 & \\
(\sim Y) & 1010 & \\
(\sim X) \wedge (\sim Y) & 1000 & \\
\sim(\sim X) \wedge (\sim Y) & 0111 & \equiv \quad (X \vee Y) \text{ (SEE 7.2b)}
\end{array}
$$

3. The input A is irrelevant, because when either B′ or the C′D combination occurs the output required by the specification formula is a 1, no matter what the value of A is. B′ occurs in eight rules. In these eight, the remaining three variables (including A) take on all possible combinations of values. There is therefore no need to test them, since, no matter what the results of the tests are, the output value is required to be a 1. The same is true for C′D and all possible combinations of values of the variables A, B.

4. The policy map is given in the following figure:

```
000000:ERROR              100000:ERROR
000001:ERROR              100001:010000100
000010:ERROR              100010:010010000
000011:ERROR              100011:ERROR
000100:ERROR              100100:100100000
000101:ERROR              100101:ERROR
000110:ERROR              100110:ERROR
000111:ERROR              100111:ERROR
001000:ERROR              101000:ERROR
001001:001000011          101001:ERROR
001010:001000011          101010:ERROR
001011:ERROR              101011:ERROR
001100:001000011          101100:ERROR
001101:ERROR              101101:ERROR
001110:ERROR              101110:ERROR
001111:ERROR              101111:ERROR
010000:ERROR              110000:ERROR
010001:010000100          110001:ERROR
010010:010001000          110010:ERROR
010011:ERROR              110011:ERROR
010100:100100000          110100:ERROR
010101:ERROR              110101:ERROR
010110:ERROR              110110:ERROR
010111:ERROR              110111:ERROR
011000:ERROR              111000:ERROR
011001:ERROR              111001:ERROR
011010:ERROR              111010:ERROR
011011:ERROR              111011:ERROR
011100:ERROR              111100:ERROR
011101:ERROR              111101:ERROR
011110:ERROR              111110:ERROR
011111:ERROR              111111:ERROR
```

EXERCISE 7.4 Vintage-car limited-entry policy map

5. A policy map for figure 2.8 would require 512 squares.

Chapter 8

1. The diagram that follows corresponds to figure 3.3, but the lozenges and words have been omitted. The lozenges correspond to the positions in the diagram below in which the labels have two exit arrows. The pattern is the same as that of figure 3.3 and the correspondence should be evident.

```
XXXXXXXX→→→0XXXXXXXX:A
    ↓
1XXXXXXXX→→→10XXXXXXX→→→10XXXXX0X→→→10XXXXX00:D
    ↓           ↓            ↓
    ↓        10XXXXX1X:B   10XXXXX01:C
    ↓
11XXXXXXX→→→111XXXXXX:B
    ↓
110XXXXXX→→→1100XXXXX→→→1100XXX0X→→→1100XXX00:D
    ↓           ↓            ↓
    ↓        1100XXX1X:B   1100XXX01:C
    ↓
1101XXXXX→→→11011XXXX:B
    ↓
11010XXXX→→→110101XXX:B
    ↓
110100XXX→→→1101001XX→→→11010010X→→→110100100:D
    ↓           ↓            ↓
    ↓        11010011X:B   110100101:C
    ↓
1101000XX:B
```

This form of flowchart is much more suggestive of the kind of testing that characterizes a flowchart internal to the computer than is the conventional form.

2. The same kind of conventions are employed in the following diagram. In addition, the subtable idea is used. All the branches that terminate in a similar series of tests in figure 3.3 now branch to subtable (subroutine or subprogram) SBTB1. This consists of the tests and actions common to all the branches that lead to it.

```
XXXXXXXX→→→0XXXXXXXX:A
    ↓
1XXXXXXXX→→→10XXXXXXX→→→SBTB1
    ↓
11XXXXXXX→→→111XXXXXX:B
    ↓                                SBTB1→→→→XXXXXXX0X→→→XXXXXXX00:D
110XXXXXX→→→1100XXXXX→→→SBTB1             ↓
    ↓                                 XXXXXXX1X:B   XXXXXXX01:C
1101XXXXX→→→11011XXXX:B
    ↓
11010XXXX→→→110101XXX:B
    ↓
110100XXX→→→1101001XX→→→SBTB1
    ↓
1101000XX:B
```

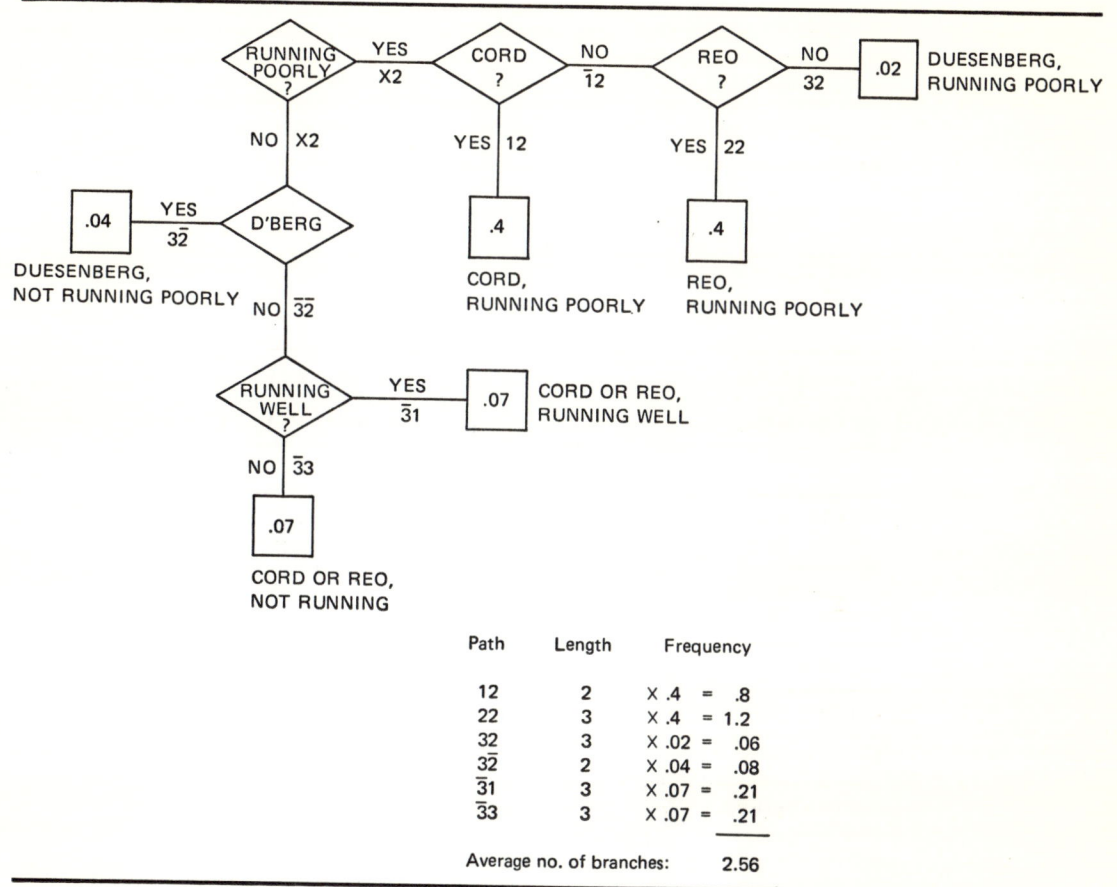

EXERCISE 8.3 Optimal flowchart for situation II
(compare with figure 8.5)

3. The diagram is identified as "Exercise 8.3   Optimal flowchart for situation II."

4. The answer is obviously 8.1. The shortest path (for Duesenberg) is of length 1. The others are all of length 2. The other diagrams all contain longer paths.

5. The diagram is identified as "Exercise 8.5   Payroll decision table in subtable form."

NW . Holiday not worked   PW : Payroll week   WS : Straight-time hours
W : Holiday worked   CW : Consecutive 7 days WSO : Time-and-a-half hours
NH : Not a holiday   TH : Time-card hours   WSD : Double-time hours

	1	2	3	4	5	6	7	8	9	11	12
Subtables	IN									A	
Holiday (NW, W, NH)	1	2	3	3	3	3	3	3	3		
5 PW days worked?			1	0	0	0	0	0	0		
5 Cons CW days worked?				1	0	0	0	0	0		
(WS + TH) > 40?					0	0	1	1	1		
TH > 8?					0	1	0	1	1		
(WS + TH − 40) > TH − 8?								0	1		
Seventh day?										0	1
Add TH to WSD		X									
Add 8 to WS	X	X				X					
Add TH to WSO			X	X							
Add TH to WS						X					
Add TH − 8 to WSO							X	X			
Add WS + TH − 40 to WSO								X	X		
Set WS to 40								X	X	X	
Calculate pay											1
Zero WS, WSO, WSD											2
Branch to	A	A	A	A	A	A	A	A	A	EX	EX

IN : Entry to table   EX: Exit from table

EXERCISE 8.5 Payroll decision table in subtable form.

Index

Action, 51, 52
Action portion of a rule, 11
Action row, 11
Action sequence numbering, 135-36
Action set, 51, 53
Action stub, 8-10, 15, 51, 52
Algorithm, 39-41
And, 99
APL, 149-50
Application, 12

Base, 73
Boolean algebra, 104-7
Boole, George, 97
Branch labels, 130
Branching and subtables, 129-30
Branching structure, 41

Carroll, Lewis, 97, 107
Codebook, 12
Coding, 12
Complementary set, 69
Completeness, 14, 16, 38-39, 56, 65, 82
Complex rule, 56
Condition, 51, 52

Condition entry, 70
Condition portion of a rule, 11
Condition row, 11
Condition set, 51, 53, 70
Condition stub, 8-10, 15, 51, 52
Consistency, 38-39, 56, 65, 81, 85-87
Consistency, interrule, 81, 85-87
Consistency, intrarule, 81, 85-87
Consolidating simple rules into composite rules, 53-55
Correspondence, 50, 51

Data bases, 151
Data communications, 151
Debugging, 130
Decision table
  advantages, 2
  and coding structures, 66
  and information gathering, 16
  and information systems, 18-19
  and narrative description, 9
  and symbolic logic, 98
  and systems analysis, 14-16
  body, 13
  coded, 11

comparison with flowchart and narrative description, 36–38
composite rules, 55–56
extended-entry, 10
flexible format, 17
formats and nomenclature, 7–14
header, 13
interactive translator, 149–50
limited entry, 10
mechanics of construction, 16–18
mixed-entry, 10
preprocessor, 147
translators, 2, 147
vertical, 14
Definition
by connotation, 52
by denotation, 52
De Morgan, Augustus, 113
De Morgan's theorem, 113
Digital procedure description, 41, 66, 71
Digital procedure descriptions, 162
Digitized description of condition sets, 70
Disjoint, 69
Don't-care entry, 13–14, 54, 69
Don't-care square, 20

Else rule, 74, 90, 166
Empty rule, 70
Empty set, 69
Entry, 10, 53
Entry portion of a decision table, 51
Exclusive or, 99
Exit, 13
Exit row, 13
Expanding composite rules into simple rules, 55

Fixed base, 73
Fixed radix, 73
Flowchart, 35–36
Full set, 69

Gardner, Martin, 97

Human engineering aspects of decision tables, 161–62

Implies, 99
Inclusive or, 99
Inconsistency, 85–87

Interleaving tests and actions, 133–35
Intersection, 69
Intratable branching, 135–36
Iterative tables, 135–36

Join, 69

Limited-entry table, disadvantages, 159
Logical connective, 99
Logical function, 99, 100
Logical map, 107–8
Logical map, computer analysis of, 111–12

Material implication, 99
Matrix, 56
Mechanizing logic, 101
Meet, 69
Minimum execution time, 131–33
Minimum program storage, 131–33
Minterm, 106
Mixed radix, 73

Natural-language programming, 32–34
Not, 99
Null rule, 70
Null set, 69
Number of rules in a table, 66–68
Number of rules in an else rule, 84–85

Optimizing branching structure, 130–31
Or, 99
Ordered set, 50
Overlap, 69–70
Overlapped composite rules, 56, 70–71

Policy map, 106, 108–10
Policy map, computer analysis of, 112
Procedure, 12, 39–41
Program, 39–41
Program logic, 10, 41, 55
Program specification, 38–39
Programming and systems analysis, 160–61
Programming languages in decision-table format, 148–49
Proposition, 98

Radix, 73
Recursive tables, 135-36
Redundancy, 38-39, 56, 81, 82
Removing rule overlap, 82-83
Row, 11
Rule, 11, 51, 53
Rules, 15
Rule, composite, 53-54, 70
Rule, simple, 53-54, 70

Scalar, 56
Set, 50
Set overlap, 69
Sets, 68-70
Set-subset relationships, 105
Sorites, 103
Stub, 8-10
Subprocedure, 12

Syllogism, 103
Systems analysis, 2-4, 38-39

Tabular techniques, 137
Tautology, 101
Tensor, 56
Theory of coding, 71
Time-sharing, 39
Top-down programming, 158
Truth table, 98

Union, 69
Universal set, 69
Universe, 69

Vector, 50, 56
Veitch-Karnaugh map, 108
Venn diagrams, 102-4
Venn, John, 97, 107

This book was designed by Naomi Takigawa, set in Press Roman (with decision tables in Univers and the whole garnished with Walbaum and a dash of Craw Modern) by Commart of Palo Alto, then printed and bound by R.R. Donnelley and Sons in Crawfordsville, Indiana. The manuscript was sponsored by Jack Maloney and edited by George Oudyn. The front cover is by Nancy Golub.

4567/54321